The Territories of Identity

The expedited globalized process of exchange and new forms of cultural production have transformed old established notions of identity, calling into question their conceptual foundations. This book explores the spatial and representational dimension of this phenomenon, by addressing how the reshaping of the key themes of place, architecture and memory is altering the nature, as well as our understanding of identity.

Cutting across boundaries, the book drives discussion of identity beyond the well-worn concern for its loss within a globalized context, and importantly provides links between identity, place, memory and representation in architecture. Examining a range of case studies from Australia, Germany, India, Japan, Mexico, Poland, Puerto Rico, Turkey and Singapore, as well as with contributions cutting across temporal boundaries, this volume addresses such issues as architecture technology, place and memory – critical issues in the monitoring and mapping of identity shift within a rapidly globalizing context.

With contributions from renowned authors in the field, including Nicholas Temple, Patsy Hely, Robert Brown, Liane Lefaivre, John Hendrix, Ana Souto, Fiona MacLaren, Stephen Walker, Nezar AlSayyad, Andrzej Piotrowski, Catherine Ettinger, Luz Marie Rodríguez, and Raymond Quek, this book presents fresh insights and diverse perspectives on the evolving questions of identity and globalization.

Soumyen Bandyopadhyay is Professor of Architecture and Design at Nottingham Trent University, UK.

Guillermo Garma Montiel is Principal Lecturer in Architecture at Nottingham Trent University, UK.

The Territories of Identity

Architecture in the age of evolving globalization

Edited by Soumyen Bandyopadhyay and Guillermo Garma Montiel

Routledge
Taylor & Francis Group
LONDON AND NEW YORK

First published 2013
by Routledge
4 Park Square, Milton Park, Abingdon, Oxon OX14 4RN
605 Third Avenue, New York, NY 10017

Routledge is an imprint of the Taylor & Francis Group, an informa business

British Library Cataloguing in Publication Data
A catalogue record for this book is available from the British Library

Library of Congress Cataloging in Publication Data
The territories of identity : architecture in the age of evolving globalisation / [edited by] Soumyen Bandyopadhyay and Guillermo Garma Montiel.
pages cm
Includes bibliographical references and index.
1. Architecture and globalization. 2. Culture and globalization.
I. Bandyopadhyay, Soumyen, editor of compilation.
II. Garma Montiel, Guillermo, editor of compilation.
NA2543.G46T47 2013
720.1'03—dc23
2013018202

ISBN: 978–0–415–62287–5 (hbk)
ISBN: 978–0–415–62288–2 (pbk)
ISBN: 978–1–315–88149–2 (ebk)

Typeset in Dante
by Swales & Willis Ltd, Exeter, Devon

Contents

Illustration credits

The authors and publishers would like to thank the following individuals and institutions for giving permission to reproduce material in this book. Every effort has been made to contact copyright holder for permission to reprint material. If any errors or omissions have been made we would be happy to correct them at a later printing.

Introduction:
Photographs: Soumyen Bandyopadhyay.

Chapter 1:
Drawings: Nicholas Temple: 1.1, 1.3 (drawn after archaeological reconstruction), 1.4.
Photographs: Nicholas Temple: 1.2a, 1.2b; Amanda Claridge: 1.5.

Chapter 2:
Photographs: Dixson Gallery, State Library of NSW [DG V1/14]: 2.1; Glen Duffus Photography: 2.2.

Chapter 3:
Photographs: Robert Brown: 3.1, 3.2, 3.3.

Chapter 4:
Drawings: Tom Paxton (after Bhatia, G. 1994): 4.1 (Bhatia 1994: 110), 4.2 (Bhatia 1994: 87).
Photographs: James Traynor: 4.3, 4.4, 4.5, 4.6.

Chapter 5:
Photographs: Cemal Emden: 5.1, 5.2, 5.7; Canan Dagdelen: 5.3, 5.6; Emre Ogan: 5.4; Erhan Sermet: 5.5.

Chapter 6:
Photographs: John Hendrix: 6.1, 6.2; Madison Digital Image Database: 6.3; Tom Sherman: 6.4.

Chapter 7:
Photographs: Ana Souto: 7.1, 7.2, 7.3, 7.4, 7.5.

Chapter 8:
Drawing: Guillermo Garma Montiel (after Enrique Espinosa López 2007): 8.1.
Photographs: Guillemo Garma Montiel: 8.2, 8.3, 8.4, 8.6; Ernesto Muñiz: 8.5.

Chapter 10:
Photographs: Leeds Museums & Galleries (Henry Moore Institute Archive) and The Helen Chadwick Estate.

Chapter 12:
Photographs: Andrzej Piotrowski: 12.1, 12.2, 12.3, 12.4.

Chapter 14:
Photographs and drawings: Frederick Seelman/Robert Swartburg: 14.1 (Architects' Association Miscellaneous Projects Collection/Architecture and Construction Archives, University of Puerto Rico); Henry Klumb Collection (Architecture and Construction Archives): 14.2; Toro-Ferrer Collection (Puerto Rico Historic Buildings Drawing Society) /Andy Rivera: 14.3.

Chapter 15:
Photographs: Royal Institute of British Architects: 15.1a, 15.4; Raymond Quek: 15.1b, 15.2, 15.5; Dr Ho Kok Hoe: 15.3.
Composite image: Raymond Quek collection, except RIBA: RUMAH 1961: 15.6.

Contributors

Nezar AlSayyad is Professor of Architecture, Planning, Urban Design and Urban History at the University of California, Berkeley. He is a widely published author and the Chair of the Center for Middle Eastern Studies, the Director of the International Association for the Study of Traditional Environments (IASTE) and the editor of *Traditional Dwellings and Settlements Review* (TDSR). His publications include *Cairo: Histories of a City* (Harvard University Press 2011) and *The Fundamentalist City?* (Routledge 2010).

Soumyen Bandyopadhyay is Professor of Architecture and Design at Nottingham Trent University. He has researched and published widely on aspects of Indian modernity and traditional Omani architecture. His recent publications include *Manah: Omani Oasis, Arabian Legacy: Architecture and Social History of an Omani Oasis Settlement* (Liverpool University Press 2011) and *The Humanities in Architectural Design* (Routledge 2010, edited with Temple, Lomholt and Tobe).

Robert Brown is Head of Architecture at Plymouth University. He has various publications on socio-cultural identity and architecture, including chapters or articles in *The Handbook of Architectural Theory*, *Traditional Dwellings and Settlements Review*, *Engaging in Architectural Education*, and *Transactions*. His research interests also include urban design methodologies and architectural pedagogy.

Catherine Ettinger is an architect with a doctorate from the UNAM in Mexico City. She currently teaches and conducts research at the Universidad Michoacana de San Nicolás de Hidalgo in western Mexico. Her research interests include architectural theory and modern architecture, in particular the flow of ideas between the United States and Mexico.

Guillermo Garma Montiel is Principal Lecturer in Architecture at Nottingham Trent University. He trained as an architect in Mexico City and arrived in the UK in 2002 to carry out postgraduate studies at the University of Nottingham. He is currently undertaking his Ph.D. at Nottingham Trent University, which discusses the notions of identity and globalization in contemporary Mexican architecture.

Patsy Hely is an academic and a practising artist. She completed her Ph.D. 'Ceramic Objects and the Articulation of Place' at Australian National University (ANU) in 2007. Research interests include colonial ceramics and ceramics and identity. She is currently Graduate Coursework/Honours Convenor in the School of Art at ANU.

John Hendrix is Professor of Architectural History at the University of Lincoln, UK, and a Lecturer at Roger Williams University, US. He has written several books, including *Architecture as Cosmology*, *Architecture and Psychoanalysis*, and *Architectural Forms and Philosophical Structures*. He is also a co-editor of *The Cultural Role of Architecture*.

Liane Lefaivre is the Chair of Architectural History and Theory at the University of Applied Arts in Vienna and a research associate at the Technical University of Delft. A multiple award-winning architectural historian and theoretician, she has published extensively and lectured around the world. With Alexander Tzonis, she coined the concept of Critical Regionalism. Their most recent book on the topic is *Architecture of Regionalism in the Age of Globalization* (Routledge 2012).

Fiona MacLaren is Senior Lecturer in Photography at Nottingham Trent University. Her research practice harnesses photography, sound and moving image to examine the perception and understanding of space, architecture and the politics of place. Current writing centres on the agency of memory in the critical representation of history and place.

Andrzej Piotrowski, Professor at the School of Architecture, University of Minnesota, USA, was educated as an architect in Poland and practised architecture in Poland and the United States. He is the co-editor of *The Discipline of Architecture* and the author of *Architecture of Thought* (University of Minnesota 2011). His scholarship focuses on the interconnections between representation, epistemology and design.

Raymond Quek is Professor of Architecture and Head of Discipline (Architecture) at the Institute of Sustainable Development and Architecture, Bond University, Gold Coast, Queensland, Australia. His academic research explores architectural representation in the history of ideas as a problem of knowledge and the meeting of East and West from the fourteenth to the eighteenth centuries. He is editor of the volume *Nationalism & Architecture* (Ashgate 2012) and author of the forthcoming book, *Fiction • Image • Tone: Metropolis & Architecture Witnessed* (Routledge).

Luz Marie Rodríguez holds a Ph.D. in Theory and History of Architecture from the Polytechnic University of Catalonia in Barcelona. She is Associate Dean of Architecture at the Pontifical Catholic University of Puerto Rico. Her research deals with architecture as instrument in the construction of Puerto Rico as subaltern.

Ana Souto is Senior Lecturer at Nottingham Trent University. She holds a Ph.D. in Latin American Studies from the University of Nottingham and her research interest lies in architecture as a cultural manifestation of national identity, in particular in Mexico and Spain. She is currently involved in a research project on Memory and Identity that aims to explore the vital role of museum architecture in the construction, reconstruction and/or narrative of a nation's identity.

Nicholas Temple is Professor of Architecture at the University of Huddersfield, having previously taught at institutions in the UK and USA. A Rome Scholar at the British School at Rome, he was recently awarded the Paul Mellon Rome Fellowship. Among his published works is *renovatio urbis: Architecture, Urbanism and Ceremony in the Rome of Julius II* (Routledge 2011).

Stephen Walker trained as an architect, and has practised in the UK and Spain. Since 2001 he has taught in the School of Architecture at the University of Sheffield, and has published widely in journals such as *Grey Room, The Journal of Visual Culture, Leonardo, ARQ, The Journal of Space & Culture,* and *The Journal of Architecture*. He is author of the book *Helen Chadwick: Constructing Identities Between Art and Architecture* (I.B. Tauris 2012).

Acknowledgements

The editors wish to thank Giamila Quattrone and Habib Reza for their help during the preparation of this book. Also we would like to extend our sincere thanks to Nottingham Trent University and Francesca Ford & Laura Williamson at Taylor and Francis for their invaluable support. Finally, we would like to thank our families, Franca, Leonardo, Jagori, Tina and Rishi for their continued patience and support during the preparation of this book.

Introduction

Beyond the (post)colonial: fragmented identities in the age of globalization

Soumyen Bandyopadhyay and Guillermo Garma Montiel

Today human civilization stands at a critical juncture; the old Western orders of the latter half of the twentieth century have largely been dismantled and only partially replaced by new world orders. It could be argued that Western hegemony is on the decline or at least is being matched by the growing influence of China on world affairs and the rise to prominence of the BRIC economies. Globalized capitalism has introduced its own problems: ecological disasters, the comprehensive financial crisis, dwindling oil reserves, terrorism and food shortages are converging symptoms of a single, failed global system. Western powers, in their self-appointed role as guardians of democracy, have continued to struggle to contain the growing threat of territorial and international terrorism. The extended influences of the Arab spring have failed to bring any degree of stability and certainty to the millions hoping to make a simple living in the Middle East. In many ways the world is closer than ever before, connected by the invisible webs of communication, yet it is in the throes of unprecedented fragmentation, contestation and divergence.

The expedited globalized processes of simultaneous enhanced exchange and fragmentation, and new forms of cultural production, have transformed old established notions of identity, calling into question their conceptual foundations long considered sacrosanct. The conflation of the traditional elements and agents of identity formation, as well as their intersection with factors traditionally regarded as antithetical to collective identity, have engendered an unprecedented complexity. Today, geographical space coexists with global and virtual space, and individual memory with the collective. The nature of the present globalizing culture has made it possible for traditional identities to coexist with the emerging ones, resulting in a representational dilemma never experienced before. Identity is seeking freedom from traditional historical confines, yet anxious over its possible disappearance, altogether.

Identity inevitably relies on a certain cohesiveness of commonly held values that allows a reassurance from the 'other' regarding the 'self'. Foucault's description of the utopic nature of the image produced by the mirror draws important parallels with how the 'self' regards the 'other':

> The mirror is, after all, a utopia, since it is a placeless place. In the mirror, I see myself there where I am not, in an unreal, virtual space that opens up behind the surface; I am over there, there where I am not, a sort of shadow that gives my own visibility to myself, that enables me to see myself there where I am absent: such is the utopia of the mirror.
>
> (Foucault 1986: 24)

Like the virtual image produced on the mirror, the 'self' is both present and not present in the 'other'; yet the 'self' requires the 'other' – like its image on the mirror – to reassure itself that it exists. The problem of the self and the other is also a difficult one as these are hardly mutually exclusive categories any more.

Architecture and the built environment have always played a key role in expressing socio-cultural and political identity and have also attracted more than their fair share of criticism for their apparent failure to live up to the expectations for expressing identity concerns of diverse groups. While the extent to which architecture still retains the ability to express identity in today's schizophrenic, globalized world is debatable, and its nature exposed to renewed scrutiny, its role nevertheless continues to be accepted. This book of carefully selected contributions emerging from the conference *Multiple Faces of Identity* (2009) explores the spatial and representational dimensions of this phenomenon, by addressing how the reshaping of the key themes of *place*, *memory* and *architectural representation* has altered our understanding of identity in recent times. The scope is wide-ranging, as is its geographical coverage; contributions relate to at least thirteen different countries spanning six different continents and regions of the world. Place – the focus of study in diverse disciplines, such as architecture, urbanism, geography and sociology – provides the stage-set for the enaction of our daily lives through its tangible and intangible characteristics. Memory is intimately connected with the construction of identity; it is with the memories we collect throughout our lives that we construct our sense of identity and belonging. The way identity is portrayed through and in our built environment leads us to the final theme of the book: representation.

The construction and representation of meaning through the interplay of place making, art and architecture, and collective memory have always given shape to the expressive intent and consequently have embodied identity within the built environment. While the physicality of place has provided the sense of continuity, architecture has constructed tangible identity through the manipulation of space, the employment of materials and technology, and the resultant appearance and associated experiences (spatiality, materiality and tectonic experience). Collective memory – covering the spectrum from recorded histories to myths – has inspired cultures and offered ways of evoking deeply entrenched identity concerns. To what extent these views are still sacrosanct in today's globalized world and what methodological shifts should there be in the study of identity in architecture are the concerns of this volume.

Critical regionalism

Lefaivre and Tzonis in their recent publication *Architecture of Regionalism in the Age of Globalization* (2012) have reflected on the true effects of the globalizing phenomenon. Has it indeed brought to an end or even minimized the resistance to increased interaction between regions? Has true freedom of instant accessibility and communication between regions been put in place? Has enhanced transport and communication brought to an end the disparities between the core and the periphery, the subjugation of peoples, cultures and individuals? On the contrary, there is enough evidence to support the view that it has led to heightened polarities of economic prosperity and deprivation, and of environmental opulence and destruction. 'With global "flattening"', the adversaries of globalization argue, 'the only things that got levelled were the unique peaks and valleys of ecological diversity, ways of life, and cultural treasures, along with nature, ideas and identities' (Lefaivre and Tzonis 2012: 2).

Critical regionalism has offered an important framework for understanding identity within the built environment of postcolonial countries, a need Lefaivre and Tzonis, and subsequently Frampton, had championed in the 1980s in the face of significant erosion of local building traditions brought about by industrialization and the use of standardized building materials (e.g., Tzonis and Lefaivre 1981; Frampton 1985; 1998). Lefaivre and Tzonis make a case for the continued relevance of regionalism urging that the approach continues to strive hard to maintain its Kantian criticality. While there is no reason to question the need for a critical position, especially in the light of the flattening of uniqueness indicated above, it seems that identity in architecture would be required to perform two important roles. It should not only express the social, political and cultural peculiarities of the present time, but it would also need to be an instrument for resisting the destructive effects of globalization. In this, one could detect continued traces of the modernist penchant for the high moral/ethical position and the quandary that identity studies found itself in from its very inception.

It is mainly due to the expedited erosion of local traditions in the latter half of the twentieth century – but also fuelled by a growing postmodernist and poststructuralist consciousness – that debates about 'local identity' and what constitutes 'the own and the foreign' came into prominence during the 1980s (Herrle 2008: 13). Such identity concerns arose from an acknowledgement of a deep crisis confronting the traditional foundations of cultures, mainly fuelled by the realization that modernism had failed to deliver on its redemptive promises of progress, morality and aesthetics. Today, as Herrle puts it, it is characterized by the growing 'decomposition' and disappearance of centralized institutions and collective minds as societies and cultures move through modernization and globalization (Herrle 2008: 11). The belief in certain conservative quarters that this process is reversible or that we are simply passing through a transitory phase soon to become stabilized, is erroneous and problematic. Critical positions in architectural identity discourse need to deal with this deeply unstable, volatile and even chaotic situation.

The key concerns discussed in this book – place, memory and identity representation – as this introduction argues, arise as critical regionalism emerges from the claustrophobic tussle between modernism and postmodernism. Conceived mainly as *resistant* modernism, postmodernism and its sibling, postcolonialism, were obsessed with defining themselves in opposition to the modernist and the colonial projects respectively. The polarized dualities of static categories that characterized such positioning resulted in a loss of understanding of historical continuity between the pre-modern past and the present, severing the traditional ties between history, narratives and collective memory. The dynamic, *heterotopic* nature of place and identity was lost in this opposition.

Fixed locations, static categories and flat narratives

Critical regionalism more recently has come under criticism for falling short of explaining many of the contemporary issues, especially those facing the so-called Third World countries. The principal concerns have been: its tendency to overlook historic backgrounds of countries – especially those in the southern hemisphere; its basis in largely static and mutually exclusive notions of culture and identity, and modernity and tradition; and its continued emphasis on location – the congruency of the physicality of territory with cultural domain – as the source of the identity content generated by the 'self' (Herrle & Schmitz 2009: 9). The latter position, by extension, has also struggled to deal with negotiations between the 'foreign other' and the 'self', a necessary means since time immemorial by which cultures and societies have been revitalized and identities shaped. While critical regionalism has emphasized the dynamic, 'variant and changing' 'chameleonic' character of regional identity (Lefaivre and Tzonis 2001: 4), the foreign human agency has never really been articulated. The need to better understand the role of the foreigner is now perhaps more pressing than it has ever been, not only due to the urgency to describe the identity formations in the age of rapid global transformations but also to recover the lost historic territories of identity formation, before and during the colonial periods and in the early postcolonial era.

Indications of these problems lie embedded in the very inception of critical regionalism as a strand of postmodern thinking. The project emerges from the postmodernist positions that sought to deconstruct modernism to *exceed* modernism – from a postmodernism of *resistance* (as opposed to the postmodernism that repudiated modernism for its own celebration – a postmodernism of *reaction*, evidenced through the works of Johnson, Graves and the like; for discussion see Foster 1985: ix–x). To resist, as Frampton proclaimed, architecture needed to assume an *arrière-garde* position to remove itself equally from the Enlightenment myth of progress and the conservative urge to return to the pre-industrial past for its essential resources for building the present (Frampton 1985: 20). 'The fundamental strategy of Critical Regionalism' as Frampton elaborates, 'is to mediate the impact of universal civilization with elements derived indirectly from the peculiarities of a particular place' (Frampton 1985: 21). Maintaining a high level of self-consciousness and deriving the governing inspirations from the local are emphasized in the mediation between cultures and universal civilization; criticality is to be achieved through a manifest critique of the latter. Focused on polarities and location, the pliable nature of the 'foreign other', although implicitly acknowledged by Frampton, is never really explored (see, for example, discussion of European influence on early twentieth century Californian architecture; Frampton 1985: 22).

Ironically, such unintended simplified dichotomies and static oppositions arise out of resistant postmodernism's continued belief in the traditions of modernism (Foster 1985: vii), which perhaps could be best illustrated through aspects of the postcolonial self-critiques emerging in the 1990s. The dark shadow of colonialism has continued to hang over cultural criticism of many parts of the world, including that of the Western world. The various postcolonial approaches have been criticized for their often disproportionate concentration on colonial encounters. Presenting a collection of essays on Indian history and culture Loomba and Kaul ask: 'The question that often underlines [...] exchanges is whether "post-colonialism" continues to serve a useful purpose or whether it now deflects attention from the complexity of disparate situations in "third world" societies?' (Loomba and Kaul 1994: 4). The postcolonial criticism, ironically, has focused on the colonial past, isolating colonial periods from their previous histories as well as the present conditions. Current socio-political and concomitant spatial dynamics have largely been neglected as historical studies have often focused on the present as a product of the colonial past. Also, by heavily emphasizing the colonial encounter, the unravelling of the relationship between society and culture before and during colonialism, and in the postcolonial period, has remained unsatisfactory (Loomba and Kaul 1994: 5–6).

Another important problem confronting identity studies in architecture is what Sarkar has termed the monochromatic flatness of narrations of Third World histories, accepted by both First and Third World academics alike. They are the product of simplified, singular perspectives that rely on binary oppositions and as Sarkar claims, 'owes a lot to an unstated conviction [...] that historical processes in the Third World had been without much depth or thickness' (Sarkar cited in Loomba and Kaul 1994: 11). The inertia of impoverished, commodified historical accounts – rendered invisible through their narrative flatness – assumes a false aura of stable validity, as fragments of events are brought together within an artificially completed matrix, given physical expression through architecture and urban form (for Walter Benjamin's anticipation of the random and colliding historical phenomenon, see Boyer 1996: 5). The role of history as an instrument of representation and its close relationship with collective memory – both with incredible abilities to coalesce, compress and expand time and space – have thus been severed, replaced instead with sequential linearity and defined spatio-temporal thresholds.

Emergent territory, contemporary place

Foucault's seminal text, *Of Other Spaces*, first delivered as a lecture in 1967 and published posthumously, describes the contemporary anxiety with space, which far exceeds concerns with time; the latter has now assumed a subsidiary 'distributive' role for the elements spread out in space (Foucault 1986: 23). For Foucault, it was Galileo who opened up the localized medieval space, made up of hierarchic, opposing and intersecting places – the space of *emplacement* – with the space of *extension*. The mediaeval space, once dissolved, 'a thing's place was no longer anything but a point in its movement, just as the stability of a thing was only its movement indefinitely slowed down' (Foucault 1986: 23). Today, emplacement has been replaced by *site* – defined by the proximity between points and elements. The problem of *siting* or *placement* is about apprehending the markedly greater fluidity and pliability of contemporary space; it is one of knowing what relations of proximity, what circulation and classification of human elements should be adopted to achieve certain ends. The mediaeval certainty of place has thus given way to the contemporary problem of placement. The predicament of the contemporary site is akin to a floating boat – 'a floating piece of space, a place without a place, that exists by itself' – simultaneously introverted yet given over to the infinite extension of the sea, moving from port to port, accumulating its treasures (Foucault 1986: 27). The heterotopic spaces in our contemporary culture, that Foucault describes, accumulate their complexity and richness – their identity – by such means. They are like 'counter-sites, a kind of effectively enacted utopia in which the real sites, all the other real sites that can be found within the culture, are simultaneously represented, contested, and inverted' (Foucault 1986: 24).

That the contemporary built environment could be characterized in terms of simple oppositions and explained through singular narratives is both highly problematic and impoverishing. The artificial divide between the different historical periods – precolonial/colonial/postcolonial or premodern/modern/postmodern, for example – reduces the essential continuity of human civilization into a mere taxonomical exercise and even apportions decontextualized iconicity to moments of rupture. Bhabha's notions of *hybridity*, adopted with great enthusiasm by researchers of contemporary architectural culture, have perpetuated the artificially disjointed treatment of spatial phenomenon. In *Of Other Spaces* Foucault's observations regarding the dynamic and palimpsestic nature of contemporary place – exemplified in his treatment of such diverse entities as the cemetery and the fairground – provides an important conceptual alternative that has existed since late modernism.

A charting of the development of Nek Chand's Rock Garden in Chandigarh, next to the Capitol Complex designed by Le Corbusier – the administrative and legislative hub of the newly created Indian state of Punjab – would give us an insight into the contemporary heterotopic nature of identity within art and architecture. From its surreptitious beginnings as a collection of rocks in the late 1950s, the Garden evolved in its early phase as a pariah (Phase-I) (Bandyopadhyay and Jackson 2007: 11–37). Following its alleged 'discovery' in the late 1960s, it continued in its illegal existence until 1973, when a decision was taken to save the Garden from destruction and to preserve it intact. Following its inauguration in 1976 the Garden expanded (Phase-II) with the inclusion of new areas of large group displays of sculptures clad in chromatic porcelain or glass fragments. In the more recent expansion (Phase-III) the Garden has engaged with large sculptural installations and architecture, and landscape architecture interventions. The Rock Garden's relationship with Chandigarh – a city which it now effectively represents, being the second most important Indian tourist attraction after the Taj Mahal – has at times been one of opposition and at others of collaboration.

Le Corbusier's – but also Louis Kahn's – 'outsider' influence on the development of immediate postcolonial Indian architecture is well known. Such external influences were soon internalized by the emerging generation of architects – Correa, Doshi and the like – and turned into expressions of their own. At the Capitol Complex in Chandigarh and in Ahmedabad, Le Corbusier arguably anticipated the need for a critical regionalist identity in Indian architecture that would steer clear of the uneasy juxtaposition of neoclassicism and traditional iconography attempted by Lutyens in Delhi. This he did by employing two important concepts that were simultaneously universal and appealed to the Indian cultural context of the time – the primordial *fragment* and the *unfinished* or the *incomplete* (Temple and Bandyopadhyay 2007). At the Capitol Complex, as Peter Carl notes, an archipelago of architectural fragments of immense scale and proportion rises above the *tabula rasa* – the podium on which the buildings sit (Carl 2003: 106). The incomplete nature of the buildings is not only palpable in their appearance but also in their spatial organization, which resulted in the subsequent struggles to make those fit for purpose, and even led to the eventual abandonment of some. The incomplete is also present in the contemporaneous work of Charles Correa in Ahmedabad (Gandhi Memorial Museum, 1958–63), where isolated interior space-cocoons recede, revealing the scaffold of the brick pylons and the horizontality of the floor. Le Corbusier's modernist iconographic representations of industrialized progress are replaced in Correa with representations of the rural landscape.

Nek Chand's Rock Garden – languishing in the secrecy of the wasteland neighbouring the Capitol Complex – also employed primordial forms, which were retrieved – or discovered – from the riverbed in the form of bizarrely shaped stones. Some are humanoid, others reminiscent of creatures past and present – horses and bulls, and even strangely amorphous, amphibious Jurassic ones (Figure i.). Nek Chand also collected fragments which came from the destruction of the many villages to make way for the city. Like the primeval symbols that encrust Le Corbusier's buildings, these fragmentary adornments form a constant reminder of the lost certainty and stasis of an innocent world, fast replaced by

Figure i. Rock Garden Chandigarh; an example of rocks collected by Nek Chand from the nearby riverbed.

the contemporary spaces of flow; they move out of the past and engage with the present.

Encrustation or garbing – the application of *vetêment* – assumes significance as a large number of sculptures are clad in colourful bands of broken bangles, enhancing – or even imparting life to – the sculptures (Figure ii). Turned into a multitude of empowered young women, they face the audience in their multitude; they ask questions. The chromatic encrustation of the *brise-soleil* at the High Court with the bull motif, the enamel-painted door at the Assembly, but especially the eight giant tapestries at the High Court that elevate the judges from the ordinary to the exalted, are examples of garbing that Le Corbusier had also employed (AlDallal *et al* in press). While reflecting on and reworking the foreign influences thrust at his doorstep, Nek Chand quietly developed his place as an 'outsider artist' – precariously perched on the threshold of India's increasingly formalized and institutionalized modern art.

The emergent subaltern 'otherness' of the Rock Garden had shown astonishing ability to collect and absorb wide-ranging influences, which were later playfully yet painstakingly reworked to carve out its own identity through

re ii. Rock Garden, ndigarh; bangle-clad p of sculptures.

resilience. The Garden is barely recognized for its authenticity and Nek Chand for his creativity; therein, however, lies the postmodern textual, allegorical, contingent quality of his 'creation'. Sculptural traditions of Buddhist rock-cut temples fuse with Mughal architecture and traditions of wall decoration, producing pavilions through which its creator is now able to look beyond the geographical confines to similar sustained and passionate creations elsewhere, perhaps towards Gaudi in Barcelona. The theatricality is sustained in the conflict between the implied strength of these large-scale concrete architectural gestures and the method of using the material itself. The process of application of entire bags of cement stitched together to produce a bulbous and more 'organic' appearance – first tried out in the large landscape initiatives – is suggestive of rag doll making prevalent in Rajasthan and introduced into the Garden by Nek Chand.

As one meanders through the monumental gorges and caverns of the latest phase, the obsession with obsolescence and redundancy becomes all-pervasive and assumes monumental, urban proportions. Stepping outside of the immediate context of a ruptured local history, the architecture is decidedly sedate, losing itself in the lacunae of romanticism that coloured the perceptions of the nineteenth century colonial painters, such as the Daniells. The architecture and landscape are unequivocally theatrical, reminiscent of settings of Bollywood films of the time, basking in their growing popularity as a tourist destination, but also reeling under it (Figure iii). This is attested by the many inscriptions of love testimonials left behind by visitors on the stone and ceramic walls of the Garden. It is this augmented status amongst the Indian populace, far outshining that of its host city, that allows the Garden the confidence to usurp the city itself, ingesting its iconic buildings and urban features into representations in *bas relief*.

It is perhaps no wonder that Nek Chand's explorations began by constructing rafts on which he sailed across the newly excavated Sukhna Lake in Chandigarh. Like Foucault's boat, Nek Chand's Rock Garden has moved from one cultural port to another to accumulate its eclectic wealth. The Garden began life in a heterotopic wasteland – a cemetery connected both to the rural landscape of the past and the founding of the new city. Here lay buried body fragments of the 27 demolished villages, and the debris from the overly enthusiastic industrial production during the city's development. The cemetery was soon added to as Nek Chand's collection of stones and objects grew; all of a sudden, the fragments them-

Figure iii. Rock Garden, Chandigarh architectural follies.

selves began assuming a degree of preciousness beyond their collective significance as the reverse of the city's development. As the Garden's existence came to light and its legitimacy was questioned, a crisis heterotopia began taking shape – a kind of quarantined restriction and a reminder to all of the perils of transgressing the city's edict. Once legality was restored, the Garden evolved fast in a true oriental fashion, resulting in the juxtaposition of several, divergent spaces, conflicting in their architecture (e.g., small to large scale) and collection (e.g., creatures to humanoids). It is also a heterochrony, in that through its collections and representations it is both a space that indefinitely accumulates time in the manner of museums and at the same time is associated with fleeting, transitory time, given its fairground-like quality through which both time and people flow. The 'organic' and *ad hoc* nature of the Garden – effectively a collection of 'rooms', where the visitor is subjected to subordination via ritualized entry and constant surveillance – stands in contrast to the organized, hierarchical city outside its walls.

The book chapters

Arising from the concerns outlined above, the three key themes that have helped us to structure the book are *place, memory* and *representation*. Place has been traditionally linked with several disciplines, including architecture, urbanism, geography and sociology. Cutting across such disciplinary boundaries, and bearing in mind Foucault's view on the shift from mediaeval *place* to the contemporary *site*, the chapters in the first part of the book revolve around two main considerations. The first is how the geographically constructed ideas of place and boundaries that shaped identity have evolved and what continuities might it hold; and the second is the diverse actors and agencies that are effecting and shaping identity today. Part two deals with the role of memory in contemporary identity formation; it begins by considering the processes through which cultural identity becomes embedded in the human psyche, and the manner in which collective memory is constructed. The authors in this section are interested in postconflict memory landscapes exemplified through art and new forms of memory receptacles, as well as the contested territory of overlapping histories and collective memories. The opportunity of representation as a field of negotiation, in which political intentions retain their ambiguous plurality, is discussed in Part three, the final section of the volume.

Temple's chapter considers the shifting relationships between perceptions of geography, empire and identity in the Greco-Roman world, and how these informed the ritual and architecture of imperial Rome. Employing Edward Said's argument that identity grew out of a 'geographically conceived world' which, as Temple argues, largely retains its validity in the modern world, the focus of the study is on the geographical and cultural divide between Rome, the Mediterranean, and those regions north of the Danube (Germania and Dacia) that were associated with 'barbarian' existence. In this context, Temple investigates the Greek concept of *oecumene* – relating to the 'known' world – that was to serve as a model of Rome's expanding geographical frontiers. Extending the debate on territorial frontiers as sites of negotiation, Hely explores the evolution of Bennelong Point on the edge of Sydney Harbour, where the Sydney Opera House now stands. A small mud-brick hut constructed between 1790 and 1795 was built by the newly arrived Governor, Arthur Phillip, at the request of a local man, Bennelong. The house, as Hely argues, continues to feature in narratives about the colonization of Australia. It provides a marker for how indigenous and non-indigenous identities were to be negotiated in future, an aspect Hely also explores in its gigantic replacement, the Opera House. Brown challenges the modernist and postmodernist contentions regarding the dislocation of the traditional place and the diffusion of its boundaries (emplacement and siting), arguing that the act of placing one's self in the world is still central to the human condition. Drawing upon the act of entry into the Japanese home, *shikii wo matagu*, this chapter examines the values this ritual emplaces in the built environment. Brown goes on to show that the ritual retains its contemporary relevance in a fast-changing Japanese society through conceptualization of space, spatial appropriation through the body and renewed ritualized practices. Bandyopadhyay's essay, through the study of the British-born architect Laurie Baker's work from the southern Indian state of Kerala, argues for a recessive role of technology in expressing active identity. Technology, with its origin in the Greek word *techne*, indicates its fundamental yet *recessive* role in constructing meaningful space and architecture. This latency of technology, by receding away from tectonic expressive determinism, makes ground for the expression of active cultural meaning. The role of technology in Baker's work, like his own 'foreignness' – latent yet selectively expressive – acted as a hinge engaged in continual mediation between India's past and present. Lefaivre discusses the work of the artist/academic Canan Dagdelen, born in Istanbul and now living in Vienna, but operating within both cultures. A global foreigner, Dagdelen's work is fascinated with the relationship between spatial memory and terrible loss. Many of Dagdelen's works explore the close relationship between space, memory and what could be termed 'effective form' in her sculptures, which requires the viewer to connect the disjointed elements, but also affects the perception of the space that surrounds it.

Hendrix argues that in order to understand the role that architecture plays in changing concepts of cultural identity, it is necessary to understand better the role that cultural identity plays in the human psyche. Cultural identity, Hendrix argues, is configured by language – a shared linguistic matrix is what underpins the unconscious. He proceeds to reflect on how the physical and sensual experience of architecture can be absorbed into its symbolic and linguistic framework, allowing architecture to continue to resonate in a culturally pluralistic world. Cultural associations of particular elements of the architectural vocabulary raise questions about cultural specificity and identity formation. Souto observes the emergence of new considerations regarding the relationship between architecture and memory in the twenty-first century – especially in the form of museums, memorials and traces – a debate that can be seen to permeate both the new literature on memory and the built environment. Following the reunification of Germany, a renewed debate involving the question of collective memory, history and the problem of portrayal of memory through architecture has surfaced. Approaching the problem from a phenomenological position, Souto analyses two sites of memory in Berlin: Topography of Terror and the Memorial to the Murdered Jews of Europe. Garma Montiel's essay starts from the premise that cities and their architectures are material manifestations of intertwined meanings, memories and values, which play a pivotal role in the construction of our sense of identity and belonging. His chapter centres on Mexico City, and discusses the connections between the country's history, its collective memories and ideals embodied in its architecture by using two examples – La Plaza de la Constitución and the Temple of Santa Teresa la Antigua. These emphasize how the meanings and memories of the past have had a powerful influence on the construction of the present, creating a fragmentary identity, which uses architecture to adjust to the demands of the globalized capitalist world. MacLaren explores how memory forms both the central premise and the instrument in the relationship of identity, place and representation in postconflict Northern Ireland. Here, memory is appropriated in the representation and spatialization of historical experience. Incessantly claimed and contested, the dichotomy of remembering and forgetting becomes manifest, spatially. MacLaren's chapter posits the politics of place as central to the negotiation of identity, belonging and the representation of plural history, and argues the critical role of visual and spatial practices in the processes of postconflict transformation. Walker's essay on the British artist Helen Chadwick introduces us into the overlapping world of collective and personal memories. He examines four key areas of thought and exploration by Chadwick around these issues, involving her early built critique, her conceptual expansion concerning the relationships between built environment and identity, the role of the body and the impact of scientific thinking on its deployment as metaphor, and a call for ambivalence in the conception and actualization of our relationships with ourselves, with others, and with the spaces around us.

AlSayyad's paper argues that the twentieth century witnessed the eventual dissolution of specific urban forms that accompanied the transformation of political relations between the First and Third Worlds. The era of colonialism witnessed the emergence of specific hybrid forms that captured the complex relationship between the dominant and the dominated. The same modernist ideologies were often invoked to rebel against colonial order during the independence movements that emerged. AlSayyad argues that the era of globalization has introduced pseudo-modern and postmodern forms, divorced entirely from the context of place, which cast doubt on urbanism's ability to represent the cultures in which it exists. For Piotrowski, the symbolic functioning of buildings is anything but unequivocal, which highlights the limitations inherent in the current logo-centric models of cultural studies. His essay, using the Holy Trinity Chapel in Lublin, Poland, as an example, argues that many constructed spaces do not project conclusive messages, and yet they have, over time, played an instrumental role in shaping cultural identity, supporting highly nuanced non-verbal cultural exchanges. A field of representational negotiations is created, where tensions between different religions and worldviews have been made visible to heighten cultural awareness and engender unorthodox considerations. Ettinger's essay is devoted to the portrayal of Mexican architecture in the American and Mexican press, which reveals that while from within Mexico there was a great desire to showcase the modern, a different image was forged from without that emphasized tradition and history. By the 1980s the diversity of Mexican architecture – both inside and outside – had been reduced to a myopic take on tradition and superficial understanding of the role of colour, an image strengthened further by the Pritzker award for Barragán in 1980. Ettinger argues that the explanation lies in the foreigner's desire for heterogeneity in the face of homogenizing processes. Within the country, it is related at least in part to the continued acceptance of the viewpoint of the foreigner. Rodriguez's study highlights the support modern discourse and architectural aesthetics received in Latin America through its political and cultural institutions and constructs. In Puerto Rico nationalism was a problematic concept; an international aesthetic was thus partly adopted to avoid the representation of Puerto Rico as a nation. Approaching the problem from a postcolonial perspective, Rodriguez argues that the Caribe Hilton hotel (1946–1949) appears to have been pivotal in the political and cultural negotiations between the island and the United States. Within the scope of 'colonial mimicry' advanced by Homi Bhabha she argues that internationalism worked as the equalizer to the metropolis, while tropicalism was another mechanism to [re]create an exploited image of difference. Quek's historical narration reflects on aspects of Singapore's national effort towards modernity in the architectural sphere, and focuses on its curious character in publications – the printed mirror of colonialism, nationalism and independence as seen in the chronicles of early Singapore architectural journals. The first architectural journal in Singapore, which dates back to 1929, managed faithfully to sustain its benign colonial mentality until the war. It was re-established in 1951; this time the concerns shifted towards the specificity of the tropics and, in particular, nationalist sentiments. The fever of nationalism is reflected in the competition announced in the journal in 1960 for the now forlorn National Theatre in Singapore. The irony is that even by the time it was built in 1962, Singapore was not an independent nation state, which happened only in 1965.

Part One

Place and identity

Chapter 1

Changing identities in the *Oecumene*

Geography and architecture in the Greco-Roman world

Nicholas Temple

Introduction

In his seminal text *Culture and Imperialism*, Edward Said makes the following claim: '[. . .] it is the case that no identity can ever exist by itself and without an array of opposites, negatives, oppositions: Greeks always require barbarians, and Europeans Africans, Orientals [. . .]' (Said 1993: 60).

In his argument constructs of identity inevitably require judgements about otherness and difference, whereby questions of authority, freedom and justice are circumscribed by particular ideological positions. These positions are most palpably manifested in military and political actions, in particular in the conquest and annexing of land by a foreign force. Indeed, according to Said the very concept of identity, and its associations with forms of hegemonic rule, has historically evolved through the agency of a 'geographically conceived world' (Said 1993: 61).

> The actual geographical possession of land is what empire in the final analysis is all about. At the moment when the coincidence occurs between real control and power, the idea of what a given place was (could be, might become), and an actual place – at that moment the struggle for empire is launched [. . .]. Imperialism and the culture associated with it affirm both the primacy of geography and an ideology about control of territory. The geographical sense makes projections – imaginative, cartographic, military, economic, historical, or in a general sense cultural. It also makes possible the construction of various kinds of knowledge, all of them in one way or another dependent upon the perceived character and destiny of a particular geography.
>
> (Said 1993: 93)

Whilst the idea of geography – as a quantifiable and measurable parameter – is a largely modern concept, facilitated through developing techniques in surveying and cartography from the Renaissance onwards (Woodward 2007; Cosgrove 1992), the perception of land as having territorial definition, on account of its potential to be occupied, bounded and defended, has its roots in antiquity. This understanding of geography, moreover, acquired a broader strategic importance through the systems of empires. In this investigation I will explore how this strategic understanding of geography provided the basis of a highly sophisticated symbolism, which communicated difference between ordered (civilized) societies and their antitheses. At the same time, this symbolism had pretensions of universality, particularly during the Roman Empire, wherein notions of geographical boundary were circumscribed by cosmic meanings. Among the many forms that this symbolism took, the inter-relationships between architecture, ceremony and geographical terrain provided the most effective and persuasive communicative system.

Oecumene

Key to this investigation is the nature and meaning of the term *oecumene* (or *oikoumene*), as it is applied to political and ideological notions of order in antiquity. The origin of *oecumene* (from which the more modern expression 'ecumenical' derives) serves as an early indication of the emergence of geographical knowledge in the Greco-Roman world. Translated, it means 'inhabited land' and was originally used to designate the many scattered regions occupied by Greek colonies in the Eastern Mediterranean. Underlying *oecumene* however is an ethno-centred view of order in which language, religion and customs were deemed to reaffirm the ancient Greek communities as the fulcrum of the inhabited world, and the Greek way of life as an exemplar of a civilized society. Accordingly, implicit in the meaning of *oecumene* is a bias towards Greek culture as the only truly 'inhabited' community in the known world.

The location of these Greek communities, in relation to the larger landmass of Eurasia, acquires an unprecedented importance in the minds of ancient thinkers, setting in place a new geographical consciousness that would anticipate the rise of Rome and her empire. It precipitated the idea of marginal conditions of inhabitation; those parts of the world occupied by 'sub-human' beings and supernatural forces that pervade Greek myth and folklore. It was on the basis of this model of *oecumene* that ancient Hellas cultivated the persistent belief in the non-Greek as a barbarian who dwelt outside the zone of the recognized inhabited world:

> [*barbarous*] [. . .] assumed a fierce and strident ethnocentrism directed against alien folk, who differed from the Greeks in their lack of appreciation for the *polis*, the Greek language, and the literary and artistic ideals of the city-state. Greek poets, dramatists, and philosophers usually considered Greek civilization as the norm, and were fond of portraying the barbarian as the oaf, the slave, and the predator.
>
> (Jones 1971: 378–379)

Closely associated with this concept of the barbarian were certain negative attributes and a geographical locale, generally outside the Mediterranean basin:

> For the Greeks first, and then imperial Romans, 'barbarian' carried huge connotations of inferiority, in everything from morals to table manners. It meant the opposite of the 'other', the mirror image of the civilized imperial Mediterranean which the Roman Empire united.
>
> (Heather 2009: xiv)

To be uncivilized also assumed certain alien characteristics as Anthony Grafton indicates: 'The Greek writers liked to imagine bizarre beings in the hinterlands of the inhabited world, and took the ingredients for their strange portraits of strange men from many sources, from Indian epic to Greek humor' (Grafton 1995: 48). At the same time, '[. . .] [the Greeks] who invented the term *barbarian* showed that its application was broader [. . .] Strabo admitted that the term originally referred to foreigners who mispronounced Greek' (Grafton 1995: 137). Significantly, this broadly twofold characteristic of the barbarian – as an inhabitant of landlocked regions (those societies located north of the Alps) and a dweller ignorant of 'cultivated' languages – was to re-emerge during the Renaissance when Italian humanists promoted a Latinized model of proto-nationalism or 'Italianità' (Ilardi 1956).

The Greek model of the *oecumene* (and its associated meanings with respect to civilized and uncivilized societies) was inherited by Rome, albeit radically appropriated to convey the notion of terrain as potentially expansive and populated by peoples who spoke many tongues, rather than bounded and defined by a common language. As a model of empire, however, *oecumene* formed part of a larger constellation of imperial powers in the ancient world which Eric Voegelin calls the 'Ecumenic Age', beginning with the rise of the Persian Empire and ending with the fall of Rome under Byzantine rule (Voegelin 1974: 114).

Our understanding of the 'Ecumenic Age' during the period of the Roman Empire derives largely from the Greek statesman and historian Polybius (200–118 BCE), who witnessed first-hand the social and political upheavals of Roman territorial expansion. Dismayed and troubled by the relentless military campaigns of the Empire, Polybius asks why the self-contained 'concrete societies' of Hellas, Israel and Egypt declined, and were so easily absorbed into the territories of the Roman Empire. Largely content with the geographic limits of their inhabited terrain, even allowing for the expansion of the Greek colonies, these earlier civilizations drew upon their traditions and belief structures to maintain a certain order and continuity. The transformation from these relatively stable societies to the Ecumenic Age of empires required a new outward perspective, where assimilation of different religious and ethnic communities provided the impetus for a dynamic understanding of civilization:

> [. . .] [the Ecumene] was something that increased or diminished correlative with the expansion or contraction of imperial power; and it furthermore had degrees of intensity correlative with the degrees of

direct jurisdiction or indirect political control maintained from a ruling center.

(Voegelin 1974: 132)

At the same time, the Romans continued to apply the Greek model of the barbarian in their speeches and commentaries, only more clearly identified geographically through the extended frontiers of empire. Perhaps the most celebrated and influential Roman account of barbarian existence is Tacitus' *Germania*. Written in 98 AD, it describes the region of the Celtic peoples, in the area of the Rhineland. Tacitus highlights how this lost province of the Roman Empire was treated with disdain and superstition by Romans. Despite this contempt, however, Tacitus demonstrates a certain admiration for the German tribes, arguing that they are a 'primitive people, "noble savages" [. . .] endowed with desirable qualities of simplicity and uprightness that the Romans had lost' (Tacitus 1999: xxx). He distinguishes between the chasteness and moral rectitude of the peoples of Germania and 'decadent Roman morality' (Tacitus 1999: xxx). Unlike the Britons the Germans were not immigrants 'but had always been there'. Accordingly, '[. . .] the peoples of Germany have never been contaminated by intermarriage with other nations and that the race remains unique, pure, and unlike any other' (Tacitus 1999: 39).

His emphasis on the simplicity and purity of the German people is matched by their connection with the landscape, which 'either bristles with forests or festers with marshes'. These geographical features have '[. . .] made [the German people] accustomed to cold and their poor soil to hunger' (Tacitus 1999: 39). Unshaped by the oceans, ancient Germania stands outside the 'inhabited world', located geographically to the north of the defensive boarder (*limes*) that stretched from the Moselle River in the west to the estuary of the Danube River at the Black Sea to the east (Figure 1.1). Along this extensive border Roman

Figure 1.1 Map of Europe indicating th northern boundary o Roman Empire (Lime relation to the Rhine Danube rivers.

garrisons patrolled the frontier regions, including Pannonia and Moesia south of the Danube River (before Trajan's later conquest of Dacia).

The perception of Germania as being largely landlocked was due to its remoteness from the Mediterranean and the region's impregnable Teutoburgian Forest, which formed part of the *Hercynia Silva* famously described by Caesar. This forest provided an effective defence for the German tribes, an advantage demonstrated in the humiliating defeat of the Roman legions in 6 AD. Extending north of the fortified frontier, the Teutoburgian Forest became the stuff of folklore and legend, even acquiring the venerable status as the 'cradle' of German civilization by later German cultural historians (Lekan 2004: 9).

In contrast to this disorientating and threatening terrain of the north that expunged Roman military potency, the regions of the Mediterranean were considered tamer and therefore more susceptible to the civilizing effects of Roman occupation. Described by ancient commentators as the *Mare Nostrum*, the Mediterranean gave Rome its maritime dominance – but one that came at a price:

> From the end of the third century B.C. until the fall of her western empire nearly seven centuries later, Rome dominated the Mediterranean region and gradually extended her power well beyond its boundaries [. . .] Yet not even the celebrated *pax Romana* could hope to eradicate the immense diversity of provincial loyalties and cultures. There is indeed a strong sense in which the Roman Empire was not Roman [. . .] or at least was only patchily, thinly so. Rome's was an empire in which the precarious unity of Greek and Roman language and culture and an economy of exaction and coinage were totally dependent on communications; and for all the fame of the Roman road, the most basic and the most vital lines of communication lay across the sea.
>
> (Horden and Purcell 2010: 23)

The implication here is that the unity of the Roman Empire was sustained by an uncertain – even ill-defined – sense of identity among its people. This perhaps partly explains Tacitus' own longing, outlined earlier, for a return to the kind of primitive simplicity and tribal purity that characterized the Germans. Tacitus' allusion to an 'identity crisis', on account of the merging of different cultures and ethnicities under Roman rule, underlines just how important the systems of government and their associated ideological instruments were to convey a sense of order and continuity during the Roman Empire. Among the many forms that this ideology took, architecture and ceremony provided arguably the most effective means of communicating the authority of empire across such a vast geography. The symbolism deployed in these ideological instruments drew upon a complex interplay between political, military, geographical and cosmic models of *oecumene*, as the following examples will demonstrate.

Notions of centre and periphery

One example of this symbolism in action is the public ceremony of the Roman Triumph, where the everyday life of Rome was momentarily suspended and transformed into a ritualized microcosm of the whole empire (Versnel 1970; Beard 2007). Celebrating the conquest, and subsequent occupation, of foreign territories the Roman triumph entailed a procession of the victorious army through the city, culminating in a sacrifice at the Temple of Jupiter on the Capitoline. In one sense the triumphal route could be seen in symbolic terms as commemorating the *extended* geographical distance between Rome and its new frontier states. Made tangible by the display of war booty and captives, the procession would have communicated to the onlooker a sense of connecting – like an umbilical cord – the *caput mundi* with a new 'distended' periphery: 'Several ancient discussions do also highlight the role of the procession in the dramatization of imperialism and the geography of empire' (Beard 2007: 162). Polybius, for example, claimed that the triumphal march 're-presented imperial conquest at the center of the Roman world', whilst Josephus likened the 'stream of riches flowing into the city (spoils) to a river' (Beard 2007: 162). In such accounts of the Roman triumph we see geography being treated not simply as a measurable and navigable terrain but also as an ideological construct – a constellation of parts unified by a 'magnetic field' emanating from its administrative and symbolic centre (Rome).

At the same time, accounts of conquests and journeys undertaken by Roman armies were recorded in various *commentarii* by emperors and military figures, which were probably used as sources of information for the numerous sculpted scenes of battles recorded on commemorative monuments across the empire. An interesting aspect of these accounts, which has a particular bearing on this study, is the question of their transcription into maps – or *itinerarium pictum*. It is well known that the Greek world was a high point in antiquity in the understanding of geography and its representation through cartography, indicated in the celebrated studies of geography by Strabo and later Ptolemy. It seems however that this developing knowledge was not utilized in any systematic way by the Romans, who left precious little legacy in map-making (Hartley and Woodward 1987: 234). This seems odd, given the logistical challenges in governing and controlling such a vast empire, and the undoubted benefits of having reliable scaled maps for conquest and administration.

One of the few pieces of evidence that we have of map-making in the Roman world is the celebrated *Tabula Peutingeriana*, long believed to be a medieval copy of a Late Roman original. According to Richard Talbert:

> I dare to regard [the *Tabula Peutingeriana*] as having been part of a decorative scheme for a specific public space inside some imperial palace of the Tetrarchic period, when the Roman Empire was ruled by Diocletian and three co-emperors (around AD 300). The unknown mapmaker's purpose is seen as primarily artistic and celebratory, not practical or geographic; no one was ever seriously expected to plan a long

> journey or a military campaign from this map. Rather, its viewers were invited to marvel at the sweep of Roman power and civilization.
>
> (Talbert 2010: 7)

In the form of a narrow and elongated parchment scroll, covering a vast geographical territory beyond the bounds of the Roman Empire (from the Atlantic to India), the *Tabula Peutingeriana* reveals a new and unprecedented geographical reach of the ancient world. It anticipated later medieval *mappae mundi*, only in this instance Rome rather than Jerusalem is located at its centre. Besides representing key geographical features, such as rivers and mountains, and locations of towns and cities, the map also highlights roads with numerical distances indicated. This information, no doubt, was drawn directly from earlier Roman commentaries and itineraries, mentioned previously, such as the famous 'Antonine Itinerary', executed at the end of the third century (Albu 2005: 138).

The dominance of Rome on the Peutinger Map is underlined by the personification of the city as a seated and crowned figure, inscribed in a circle with shield and spear. Among the many roads shown leading into Rome, one of the most visible is the '*Via Triufalis*' (Via Triumphalis) – the road along which the triumphant armies allegedly marched before entering the city to participate in the Roman triumph. Equally prominent, in this part of the map, is the representation of Ostia, the port of Rome at the estuary of the Tiber River, which is shown in the form of a monumental curved exedra with projecting piers. This representation was intended, no doubt, to underline Rome's ancient maritime role, highlighted earlier, in its command of the Mediterranean and beyond.

As if magnifying this maritime symbolism of empire, the map also contains '[. . .] wavy lines that mark the sea coasts and also signify the vaguely identified limits of the known world,' a point that Emily Albu suggests '[. . .] serve as the decorative frame for the oecumene' (Albu 2005: 138). The significance of this aspect of the Peutinger Map draws upon an ancient belief, evidenced in Strabo's *Geographica*, that the known world was shaped by oceans. It underlines the premise that only civilized cultures were maritime societies. The idea, moreover, finds expression in ancient Greek mythology where *Oceanus*, a circumambient ocean-river, was believed to be the final destination of all rivers and streams, a belief that can be traced back to the earlier epic poems of Homer and Hesiod (Hornblower and Spawforth 1996: 1058). Located in the west, the source of the *Oceanus* lies within the inhabited world of the *Oecumene* and becomes synonymous with the Mediterranean Sea, or *Mare Nostrum*.

Trajan's Column and *Itinerarium Pictum*

An indication of the symbolic and geographical significance of ports and waterways, in the Roman understanding of empire, can be seen in the famous sculptured reliefs of Trajan's Column in Rome (Figure 1.2a). Commemorating Trajan's two campaigns in Dacia (ancient Romania), in 101–2 AD and 105–6 AD, the reliefs depict the journeys and battles of the Roman army from key territorial thresholds. More generally, the column provides an exquisite example of how the Roman Empire was understood geographically:

> Working as [the artist] did from fixed points, e.g. Drobetae, Sarmizegethusa, and Muncel Cetate, it is possible for him to indicate the topography of a campaign, not indeed with the precision of a cartographer, but at any rate within a slight, and from his point of view negligible, margin of error.
>
> (Davies 1920: 4)

It seems likely that one of the sources for the composition of the spiral relief of the column were the *Commentaries* of Trajan himself, sadly lost except for a few fragments (Salmon 1936: 90–1, n48). The influence is further testimony to the process, prevalent elsewhere in ancient Rome, of transcribing literary accounts of military campaigns into a continuous linear pictorial scene, which in the case of spiral reliefs of Trajan's Column moves chronologically from bottom to top (Coarelli 2000: 20). The transcription gives some explanation of how the Roman world communicated geographical knowledge pictorially, through an abbreviated – or compressed – chronicle of reconstructions. Attempts have been made by historians to forge connections between the locations highlighted on the *Tabula Peutingeriana* and the supposed route of the military campaigns of Trajan across the Carpathian mountains (Davies 1920: 8; Salmon 1936: 89, n34).

Figure 1.2a View of Trajan's Column.

re 1.2b View of the fs on the Column ·ajan indicating (at ɔase of the column) ɔersonification of Ɔanube and the ıe of the two bridge structions over the ·.

In the first campaign of Dacia, we are presented on the column with a scene of the Roman army crossing the Danube River, which delineated the northern frontier (*limes*) of the Roman Empire referred to earlier (Figure 1.2b). This was achieved by means of the construction of two impressive pontoon bridges, possibly intended to convey in abbreviated form two separate river crossings. In the second campaign the column shows a scene of the emperor and his army embarking on a sea voyage, probably from the Roman port of Ancona, to the Dalmatian coast en route to Dacia. By visibly connecting the Danube River, the old frontier of the Empire, with the Mediterranean coast the reliefs underline the symbolic importance of the maritime geography of the Roman *oecumene*.

Beyond the general interpretations of the column reliefs, I would like to suggest that more specific symbolic and geographical connections were intended, between the actual location of Trajan's Column in Rome and the military campaigns in Dacia. This hypothesis is prompted in part by the attribution of Apollodorus of Damascus as the architect of the Forum of Trajan and the design of a permanent bridge crossing over the Danube River (at Turnu Severin or ancient Drobetae referred to earlier). This latter construction is considered to be 'one of the most celebrated engineering feats of classical antiquity' (Salmon 1936: 95). Built some time after the first military campaign, and recorded on Trajan's Column for the 106 AD campaign, the bridge served as testimony to the permanent annexation of the territories of Dacia by Trajan, resulting in the northern extension of empire from the earlier frontier of the Danube River (Salmon 1936: 83).

Trajan's invasion of Dacia involved impressive feats of engineering that often required substantial excavations. These initiatives were commemorated in various inscriptions, such as one located on the right bank of the Danube river (west of Orsova) which reads as follows in translation:

> The Emperor Caesar, son of the divine Nerva, Nerva Trajan Augustus Germanicus, *pontifex maximus*, at the time of his fourth tribunician power, the Father of the Country, at the time of his third consulate, after cutting down mountains and removing their projecting elbows, made the road.
>
> (Rossi 1971: 49)

Called the 'Tabula Traiana', the inscription commemorates the construction of a road carved out of the side of a cliff face that extended for 65 miles to the so-called Iron Pass (Salmon 1936: 86). According to A.D. Xenopol, this remarkable piece of engineering may well have been used as a towpath for guiding ships along the river (Salmon 1936: 86, n22). A similar inscription, moreover, is found on the base of Trajan's Column (location of the later tomb of the emperor), which also refers to alterations of natural land features:

> The Senate and people of Rome to the Emperor Casear, son of the divine Nerva, Nerva Trajan Augustus Germanicus Dacicus, *pontifex maximus*, at the time of his seventeenth tribunician power, of his sixth imperial salutation, and of his sixth consulship, the Father of the Country, to declare how high a hill and place have been excavated for these great works.
>
> (Rossi 1971: 49)

As if consciously emulating the great engineering achievements of Trajan's Dacian campaigns, the height of Trajan's Column indicates the extent of excavated earth resulting from levelling 'the saddle that originally joined the Quirinal with the Capitol [. . .]' (Coarelli 2000: 4–5). Accordingly, the column acts as a kind of monumental 'surveying rod' to communicate the dramatic level changes of the site. Like the forging of roads and bridge crossings in military campaigns, the construction of Trajan's Forum similarly defies natural terrain, demonstrating in the process mastery and dominance. Consequently, conquering inhospitable mountainous regions and altering land features, for military or architectural/ceremonial ends, were it seems treated as parallel indicators of the authority of imperial rule. The point is further emphasized by the portrayal of the same mountainous terrain (the Carpathian or Cibin) in the reliefs of Trajan's Column, which provides the background to scenes of battle and military camps.

It is likely that the spiral reliefs on the shaft of the Column of Trajan were added later, after the erection of the column following Trajan's sudden death in 117 AD. Most probably commissioned by Hadrian (117–38 AD), after a special dispensation by the Roman Senate, the column became the official burial site of Trajan (Claridge 1998: 167). Whatever the initial intentions behind the design of the monument, it seems plausible that the conversion of the column base to the burial chamber of Trajan consciously drew upon the symbolism of ancient military funerary monuments. Richly embellished with reliefs of weapons of war, the exterior surfaces of Trajan's burial chamber could be compared to the famous Trophy at Adamklissi (*Tropaeum Traiani*) in Romania, built in 109 AD to commemorate the Roman victory against the Dacians at the battle of Tapae during the first campaign in 102 AD.

These geographical connections, moreover, between Rome and the actual scenes of conquest could be extended further to encompass the architecture of the whole forum (Figure 1.3). Penelope Davies argues that: '[. . .] Apollodorus based [the] unusual layout [of Trajan's Forum] upon the ground plan of a military camp such as the Praetorium of Vetera, in which the Column took the place of the army's standards' (Davies 1997: 61). In other words, the Forum serves as a 'memorialization' of Trajan's military campaign in Dacia, even a form of 'frozen re-enactment'.

A largely invisible feature of the column, at least from its exterior, is a spiral staircase that was installed within the column shaft. This passes between the levels of the burial chamber to an observatory balcony, just beneath the monumental statue of Trajan, which originally crowned the column. This statue, moreover, would probably have been visible above the roof of the nearby Basilica

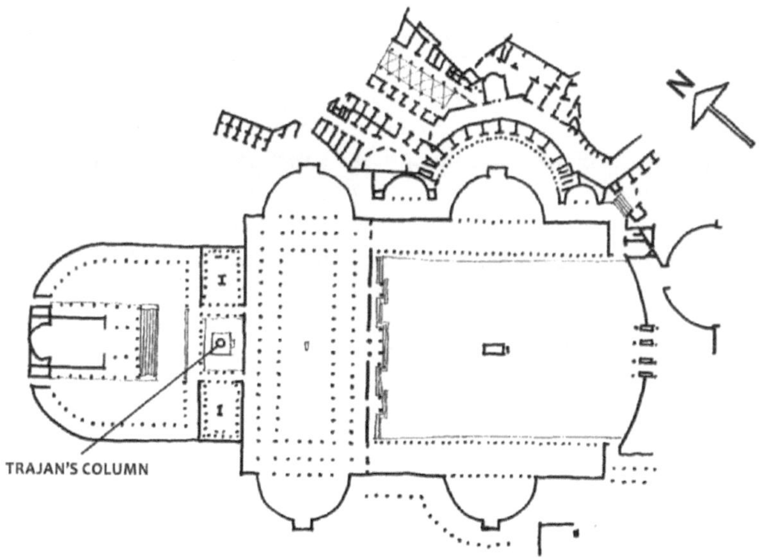

e 1.3 Plan of the n of Trajan.

Ulpia. In the journey between both levels of the column the visitor follows the course of the helix of the outer relief that rises from the mortal remains of the emperor to his immortalized representation in the colossal statue at the apex. This progression reiterates the idea of the column as an 'umbilical cord' that connects both the living/deceased and deified emperor, and at the same time links Rome with her new frontier states. Through these relationships, the monument becomes testimony to the ceremonies of *profectio/adventus* (the ritual departure and return of the emperor before and after a military campaign) and *consecratio* (his deification). As Filippo Coarelli states, through the 'faithful reproduction' of Trajan's Column as an 'altar tomb [. . .] the triumphal aspect goes hand in hand with the funerary' (Coarelli 2000: 21).

The Pantheon and *Mare Nostrum*

The Roman *oecumene* could be described as an exercise in 'world making' (*cosmopoiesis*) which, according to Voegelin, was equated with the idea of 'Pragmatic History', a history driven largely by the conquest and colonization of territory (Voegelin 1974: 133). Significantly, ancient Rome provided the first instance where cosmic symbolism was brought to the level of pragmatic history, expressed through the combination of territorial conquest, political authority and the transformations in urban topography. Whilst aspects of this symbolism were already being applied under Emperor Augustus, as highlighted for example in the triadic relationship between the Horologium, Ara Pacis and imperial mausoleum (Davies 2000: 15–16; Heslin 2007: 16), it was through the legacy of Trajan's expanded *oecumene* that this symbolism reached its most sophisticated level. The design of the Pantheon could be said to exemplify this. Attributed to the period of Hadrian, successor to Trajan, the Pantheon's vast domical space was intended to embody both the terrestrial and celestial realms, to which the pantheon of the gods variously asserted their dominion. This interpretation, which largely derives from a statement by Cassius Dio from his *History of Rome* (53.27.2), provides a useful starting point from which to examine the symbolism of the Pantheon in relation to the Roman *oecumene*.

Understood in volumetric terms, the dome forms half a sphere, the other half occupying the lower drum of the structure, with the circumference touching the slightly cambered floor (Haselberger 2009: 174; Theocharis 2009: 187–8). This twofold geometric relationship allows us to consider the interior of the Pantheon as a *mediating* space between the celestial world – defined by the 'dome of heaven' – and its terrestrial counterpart below. The alternating rectangular and circular niches, located around the circumference of the drum, probably accommodated statues of gods of the Roman pantheon, both celestial and terrestrial/chthonic (Figure 1.4). The building may also have functioned as an imperial audience hall, perhaps emulating the famous octagonal hall of the Domus Aurea of Nero described by Suetonius: '[. . .] the Pantheon provided a setting – not a temple in the conventional sense – in which the living emperor would appear in company with the gods (including [the emperor's] own predecessors)' (Claridge 1998: 206).

re 1.4 Schematic :ion through the theon.

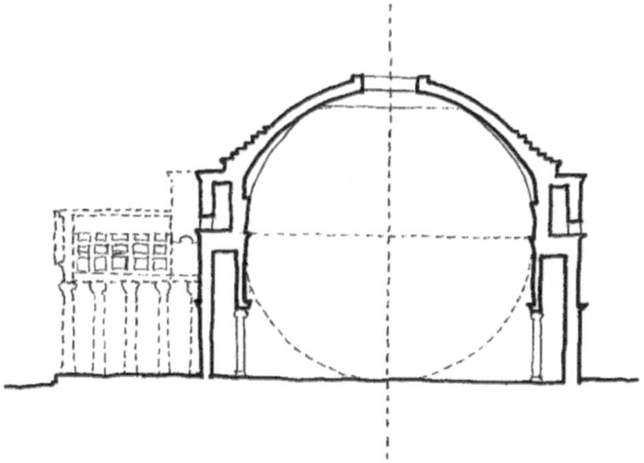

Crowning this space is the huge oculus which, it has been conjectured, marked out the seasons of the year by the passage of the sun casting light across the coffered ceiling of the dome (Figure 1.5). William MacDonald even suggests that it served as 'a kind of orrery [. . .] in which planetary implications were signalled directly as the earth revolved' (MacDonald 1976: 90).

This vast domical structure illuminates a circular paved flooring (*pavimentum*), constructed with inlaid marble (*opus sectile*) using simple geometric patterns of circles inscribed in squares. The stone used for this floor was quarried from all over the empire, a point that Patrick Hunt argues was intended to symbolize *imperium*:

e 1.5 View of aved floor *nentum*) of the eon from the s.

> The specific marbles [. . .] include purple Imperial Porphyry from Egypt (*Mons Porphyrites*), Docimian pavonazzetto from Asia Minor (*Marmor docimenium*) and yellow Giallo Numidiana marble from Carthage (*Marmor numidicum*) and – perhaps to complete a Mediterranean "square" – an as-yet unprovenanced gray Granito Grigio probably from the northwest (Pyrenees, Gaul or the Alps). Thus all four corners of the Roman Mediterranean (*Mare Nostrum*) world could be shown on the *pavimentum* floor as a squared circle as in the geography of Ptolemy, much like the Pantheon floor patterns. These *opus sectile* patterns are in repeated squares and circles that echo the overall structural geometric shapes of the building design.
>
> (Hunt 2007: unpaginated)

The stones of the *pavimentum* may well have served as *spolia* from the conquests of Egypt, Asia, Carthage and Gaul (Hunt 2007), thereby underlining the geographical reach of empire. I would like to suggest that this interpretation has a particular bearing on the Greco-Roman concept of the *oecumene* and its re-articulation in Strabo's *Geographica* and the Peutinger map. Through the material fabric of the paved floor of the Pantheon we are presented with a geological 'substrata' of the Mediterranean (*Mare Nostrum*), expressed by the juxtaposition of the various precious stones from its four regions, whose surfaces are periodically immersed in rain water cascading from the open oculus above. The floor becomes, therefore, an ideal microcosm of the terrain of empire, 'shaped' by the waters of the *Oceanus*.

The repeated motif of the squared circle, which defines the geometric arrangement of the *pavimentum*, echoes Ptolemy's geography (as Patrick Hunt argues) and at the same time symbolizes the interrelationships between the celestial (circular) and terrestrial (square) realms – the basis of the geometric order of the whole building (MacDonald 1976: 68).

The importance, moreover, of the Mediterranean in this symbolism is at least implied in the following statement:

> The claim of the Romans to 'their' sea was part of a political and cultural process by which they progressively defined the place of Rome at the heart of an Inhabited World – and *Oecumene* or *Orbis Terrarum* with the Mediterranean at its centre. This ever larger claim in the end weakened the relativism of earlier attitudes, and it was during the Roman Empire that the term 'Mediterranean Sea' itself emerged [. . .]
>
> (Horden and Purcell 2010: 12)

The Pantheon could be said to serve as an architectural expression of a 'world feeling' (MacDonald 1976: 91), or more specifically the *orbis terrarum* (earth/world) which, according to Voegelin, took on an equivalent meaning to the *oecumene* by being '[. . .] coextensive with the jurisdiction of Rome' (Voegelin 1974: 133). Considering therefore its overall configuration the Pantheon presents an idea of 'universality' in two ways: '[. . .] the circular paths of the planets under the dome of heaven and an ideated circumference of the universal empire' (MacDonald 1976: 91).

Conclusion

From these ancient perceptions and representations of geographical space we can draw a number of tentative conclusions about the concept of the *oecumeme*; firstly, that the hinterlands of Eurasia, distant from the coastal regions of the Mediterranean, were deemed formless land masses without geographical definition. Secondly, this formlessness served as an affirmation of uncivilized – or savage – communities whose existence relied upon an allegedly internal system of tribal customs (without a coherent political structure), rather than an outward-looking expansionist world-view. As Andreas Alföldi states:

> The exclusion of the peoples outside the frontier from the idea of the 'oikumene' also fundamentally affected the juridical conceptions of the late republic. What they were designating as the world was only the 'orbis Romanus' and what they meant by the 'genus humanum' was only the population of their own realm [. . .] the 'orbis terrarum' is consequently either Roman or *res nullius*.
>
> (Alföldi 1949: 5)

Through this division between the dominions of imperial Rome and *res nullius* (nobody's property) the symbolism of empire, as this chapter has sought to demonstrate, largely informed architectural renderings of the *oecumene*. Indeed, public architecture served as a crucial instrument in the Roman world for articulating interrelationships between the changing geographical limits of empire, cosmic symbolism and the cult of the emperor.

It is perhaps illuminating to consider this ancient model of cultural, political and religious order in the light of contemporary notions of geography:

> Today it is evident that the western process has produced and engulfed the global ecumene. But in contrast to widespread misconceptions and in spite of the more than 2,000 years of history that have passed since Polybius first theorized the ecumene, today's global ecumene – 'globalization' – is not a world, but is in search of a world.
>
> (Rossbach 2007: 42)

This contemporary 'search of a world' prompts us to consider the capacity of architecture today to convey a sense of *oecumene*, albeit of course fundamentally different from the Greco-Roman world-view. But is Rossbach's idea of a global *oecumene* merely a wishful ideological construct, considering the contemporary reality of increasing nationalism, ideological / religious conflict and contested territorial claims? On the other hand it is evident that the once strong historical ties of ethnic or religious communities to specific geographical locations or regions (on account of a common ancestry or tradition) are less prevalent today – particularly in the west – than ever in the past; the global diaspora has created new and emerging relationships between peoples, which are no longer simply geographically defined.

We should consider these ostensibly opposing aspects of the contemporary world in the light of Said's assertion (quoted at the beginning of this chapter) that the very concept of identity developed out of a 'geographically conceived world'. Hence the apparent demise of a collective 'geographical horizon' has inevitably eroded the historical continuities of past identities. In its place we are witnessing the emergence of forms of 'transculturation' brought about through migration and displacement, a topic that has been the subject of significant recent scholarship (Pratt 1992).

The example of imperial Rome in this transformation is, I believe, both salutary and instructive, given that the idea of a common identity of the peoples of the Mediterranean region was precarious and ill-defined (as Horden and Purcell have argued), in spite of the binding influences of Roman citizenship, bureaucratic systems of government and the imperial cult. What I have sought to demonstrate in this chapter is the effectiveness and sophistication of architecture in the political and religious arenas of ancient Rome to convincingly communicate a sense of order and continuity of empire, even when the very notion of a shared identity of citizenship was uncertain. Key to this initiative, as we have seen, was the spatial and symbolic representation of the empire's expanding geographical reach.

For architecture today, notions of centre and periphery could be described at best as ambiguous or multi-faceted, in which shifting points of reference (spatial, topographic, ideological) are mediated through often conflicting social, political and cultural outlooks. Unlike the ancient and medieval worlds, where the city was largely experienced as a unified whole (whose symbolic and geographical bearings ensured continuity with a larger cosmological order), the modern metropolis is broadly a formless amalgam of competing zones or territories – the product of precipitate urban growth – largely undefined hierarchically beyond the implementation of abstract planning zones. This formlessness, moreover, is matched by the encompassing nature of globalization itself, which assumes accessibility and visibility to all regions of the world – but without necessarily the benefit of insight into their cultural or historical differences. Through this essentially 'surface' view of the world, the contemporary city actively promotes its own vacancy – even invisibility – by serving as a scaffold for the display and promotion of other locations or destinations. One implication of this projective role of the city, to which we have all become accustomed through the agency of corporate advertising, is that the urban dweller inevitably acquires a restless expectation (or active intention) to be elsewhere – in some other part of the world (whether precipitated through work, leisure or simply dreams).

Whilst remote from the regional specificity of the Greco-Roman world-view, the idea of the contemporary city as a 'mantra' of the mobility of globalization was in some senses already anticipated in the Roman *oecumene*, where Rome (as *caput mundi* of the civilized world) became in effect an accumulation of diverse symbolic references to other regions of the empire and beyond.

Chapter 2

Buildings and the performance of identity

Patsy Hely

Introduction

In a discussion about commodities, the sociologist Henri Lefebvre wrote 'like all languages, the language of things is as useful for lying as it is for telling the truth' (Lefebvre 1991: 81), in part suggesting that the appearance of a thing, its material and shape languages, might conceal the conditions and processes that brought it into being. The idea that I extrapolate from this – that a building might lie – is perhaps fanciful but I use it rhetorically here to examine ways in which two iconic Australian buildings have refracted ideas about identity. Both have occupied the most prominent site in Sydney Harbour – Bennelong Point, a thin arm of land forming the southern side of the cove Sydney centres around.

The point, now home to the Sydney Opera House, carries the name of the indigenous Aboriginal man whose house occupied the spot for a short while, from 1790–1795.[1] The two buildings, Bennelong's small house (Figure 2.1) and the Opera House (Figure 2.2) could not be further apart in scale or technology

re 2.1 West view of ey Cove taken from ocks, at the rear e General Hospital). The illustration m an unknown . The foreground s marines'/officers' ing, the Governor's e is set in scaped grounds mid e and Bennelong's e is the small house chimney on the jutting out of the de of the harbour.

Figure 2.2 Sydney Opera House.

but both have acted to amplify the charge of their location, a location already resonant through association with the founding of Australia as a British colony. At various times since 1790, both have been used to construct ideas about indigenous and non-indigenous identity.

As in many countries with a history of colonization, relations between the original inhabitants of Australia and the broad colonial enterprise have been uneasy. There are multiple concerns at stake but ideas about identity run through much that has been negotiated and debated since the British stepped ashore in 1788 at what was to soon become Bennelong Point.

Two key events in the last 20 years underline this. Firstly, the Mabo Decision (1992) in which the High Court of Australia overturned the previously held concept of terra nullius – that the land was empty at the time of British settlement – was a catalyst for much discussion about relationships between indigenous land custodianship and identity.

Secondly the far-reaching exploration by the Australian Human Rights Commission (1997) into the practice of removing Aboriginal and mixed-race children from their parents and bringing them up outside their culture, a practice carried out on a nation-wide scale from the mid-nineteenth to the mid-twentieth century. The report from the enquiry, *Bringing them Home: The 'Stolen Children' Report* (Australian Human Rights Commission 1997), documented, among other things, the repercussions of this practice for children's, parents' and families' identities.

These were both monumental enquiries with far-reaching ramifications; they resonate and continue to be debated today. A lesson I take from both here is that the constant iterations of the past produced by historical processes engender particular constructions of identity. In this chapter then, I look at the ways in which Bennelong's small brick house has been conceived variously through the historical record, and at the changing ways the Sydney Opera House has written it into its own complex story.

In the process, I question whether built structures have the power to speak about themselves, revisit eyewitness accounts of the building of the house to clarify the context in which it came into being and the conditions under which it was built, then examine ways in which foundational stories are retold to meet particular needs and to produce particular identities.

Producing narratives

The geographer Yi Fu Tuan has commented that 'Architectural space – a house, a temple, or a city – is a microcosm possessing a lucidity that natural features lack' (Tuan 1991: 100), suggesting that planned and built spaces create a particular awareness:

> It is true that even without architectural form, people are able to sense the difference between interior and exterior, closed and open, darkness and light, private and public. But this kind of knowing is inchoate. Architectural space – even a simple hut surrounded by cleared ground – can define such sensations and render them vivid. Another influence is this: the built environment clarifies social roles and relations. People know better who they are and how they ought to behave when the arena is humanly designed rather than nature's raw stage [. . .] architecture is a key to comprehending reality.
>
> (Tuan 1991: 102)

But just as the built environment might 'clarify social roles and relations', so too do the symbolic acts buildings are attended by – especially significant ones – and the processes by which an understanding of a building's history is knitted into the fabric of public life by individuals and institutions. In looking for a way to describe such processes, Tony Bennett, in *The Birth of the Museum: History, Theory, Politics*, adopts the term 'the public historical sphere':

> [. . .] in the British context, Michael Bommes and Patrick Wright have proposed the useful term 'public historical sphere' to refer to those institutions – from museums through national heritage sites to television historical dramas and documentaries – involved in producing and circulating meanings about the past.
>
> (Bennett 1995: 132)

Australia has been in British hands since 1788, when it was incrementally taken possession of, arguably disenfranchising – on the whole – the indigenous Aboriginal population in the process. Relations between the two broad cultural groups have been in negotiation since, inflected often by guilt, and with anger, sadness and humiliation – sometimes good cheer – on both sides; a sometimes harmonious, sometimes exasperated, sometimes disinterested state of ease and unease.

The account of the small brick house that Bennelong requested be built for him is one of the nation's founding stories. Bennelong was a young man at the time the British landed at Sydney Cove, a Wangal man from one of the clans who make up the population of the Sydney (Warrane) area, collectively known as 'Eora', normally taken to mean 'of this place'. Understandings of how relationships between the two groups – the British and the indigenous population – unfolded have been the subject of much research though, with interpretations sourced primarily from early British eyewitness accounts and records.

Forming relationships

After the British, led by Governor Arthur Phillip, landed in what is now the city of Sydney, an initial period of interaction and exchange between both cultures, sometimes friendly but increasingly not, was recorded and before long the indigenous occupants had retreated to more distant parts of the harbour. The new arrivals had rapidly cleared Sydney Cove of much of its vegetation, a Georgian-style Government House had been built, fences and boundaries drawn, food gardens planted, storage facilities started, all dotted around by the wattle and daub or other rough bark shelters and tents used to house the new population. Although the numbers of indigenous occupants living at Sydney Cove in 1788 is not known, around 1,500 new arrivals with their saws, axes, shovels and brick moulds, their fishing nets and guns, must have had a great impact on the available resources.

Having failed to establish relations with the indigenous occupants by the end of 1788, Governor Phillip had one aboriginal held captive. After this prisoner died of smallpox, Phillip had two further Aboriginals kidnapped, one of whom was Bennelong. The motives spelled out were clear; Phillip wanted to learn about the indigenous population and gain information about the country and for this an interlocutor was needed (Collins 1798–1802: 113). The early eyewitness accounts suggest that Bennelong and Phillip got along well, that Bennelong was sociable and entertaining and took part in life at Government House; but he eventually escaped, which underscores of course that he was there against his will.

Then in September 1790, while visiting Manly Cove, close to the site of the original kidnapping, and attempting to make contact with Bennelong amongst a large group of his compatriots, Phillip survived a spear attack. This incident – organized by Bennelong though carried out by another – has been interpreted by some as a strategic ritual payback spearing for wrongs committed by the British.[2] Others have interpreted it as an act prompted by fear of further kidnap or of the weapons Phillip carried (Jones 2007: 42). Either way, it marked a turning point in relations. After this Bennelong and others of his compatriots were induced to come to Sydney Cove again, this time as visitors.

It was now that Bennelong asked Phillip to build him a house; a strategic act perhaps, for in 1790 only the highest-ranking British officials had even the most basic of houses to live in and living conditions by all accounts were dire. So although basic, the house would have represented a significant cost in labour and resources to the colonial force. A brick house at this time would have conferred a great deal of status on its owner, though whether Bennelong was hoping to impress the colonists or his compatriots is not clear. That there were strategic manoeuvres and negotiations on both sides is clear; and, as Nicholas Thomas has written about colonial encounters in general:

> [. . .] local relations and representations are never totally encompassed or determined by the violence of colonialism, and the distinctive forms of indigenous sociality and politics contribute in a crucial way to the dynamics of accommodation and resistance constitutive of colonial history.
>
> (Thomas 1991: 36)

As someone to make overtures toward, Bennelong was an obvious choice. He was familiar to the British, they had found him easy to get along with and he had learnt some English during his period of incarceration; he probably seemed, to a degree, adaptable to foreign ways. It could be that the British also perceived in Bennelong a desire to boost his own standing amongst his fellow countrymen and women by gaining direct access to British consumables and technologies. Believing Bennelong to be a figure of some importance, the British perhaps used him to show what might be offered to all, though there is no evidence that houses were provided for anyone else, nor that anyone else asked for one, or could make such a wish known.

From earth/land to 'thing'

The site where clay was dug and bricks for the house formed and fired, then called simply 'the brickfields', now close to the site of Sydney's main railway station, was around 2km from Bennelong Point over what was then hilly terrain. For a house of that size, based on the unusually detailed records kept of seemingly mundane details (Collins 1798–1802: 278), making five trips a day it would have taken 12 convicts approximately 15 days to pull the needed 75 cartloads of 350 bricks from the brickfields to the Point. Added to this is the digging of clay, the forming and firing of bricks, and the construction of the house itself.

It is quite possible that the British were led to the source of clay used for the bricks by the locals. In other contexts Arthur Phillip speaks of putting to use 'the clay the natives mark themselves with' (Phillip 1788) and so observing this practice perhaps led the British to the source. How the local people must have felt to see the rapid escalation of mining taking place, the opening up of huge holes and crevices, the piles of broken bricks, the whole rapidly expanding industrial conglomeration, is hard to imagine. But for a culture with a strong attachment to land, this wholesale plunder of what they surely thought was their resource must have aroused some emotion.

Why Bennelong specified the place soon to be known as Bennelong's Point for the house is not known. It could have been that it was the only one of the three most obviously well-placed sites still vacant in 1790; an observatory had been built on Dawes Point on the western arm of the cove, and Phillip's

Government House commanded the elevated mid-point of the cove. It could have been, alternately, that he recognized in the Point, in an early echo of Lucy Grace Ellem's words regarding the choice of site for the Opera House, 'the essential characteristic of a status symbol: maximum display' (Ellem 1980: 193).

The site at that time would have been quite isolated – no other buildings were near – but it had easy access between land and water and was a good vantage point for observing all that occurred in any direction. At least in claiming the remaining most advantageous spot, and one that could most likely have been seen from anywhere within the settlement, Bennelong was asserting the right not to be overlooked. Though built primarily for Bennelong, the brick structure also provided a locus for the indigenous people. Here, in a thing the British understood – a house – they could gather, they could look out, and presumably, they could be observed. And, even if they did not all make use of it, Bennelong's fellow countrymen and women would get to see and hear of its advantage; a model home in that sense, perhaps, or, as Keith Smith has suggested, 'the first Aboriginal embassy' (Smith 2001: 69).

Once built, the house was used by Bennelong and his fellow countrymen and women, according to the early accounts, primarily as a gathering place until, with a fellow aboriginal, Yemmerrawanne, Bennelong accompanied Governor Phillip to England when he returned home in 1792. After he left, his house was little used, fell into disrepair and was pulled down after he eventually returned in 1795.

Gift or exchange?

The house is documented, usually in very similar words, in all of the accounts of the first years of the colony and while the language used sometimes casts it in terms of a gift, in reality it was bound up in a whole network of exchange taking place at the time. A much-quoted chronicler, the Judge Advocate at the time, Lieutenant-Governor David Collins, recorded:

> September 1790 [. . .] at length Bennillong [sic] solicited the governor to build him a hut at the extremity of the eastern point of the cove. This the governor, who was very desirous of preserving the friendly intercourse which seemed to have taken place, readily promised [. . .]
>
> (Collins 1798–1802: 113)

And another eyewitness account, that of Captain-Lieutenant of Marines Watkin Tench:

> Baneelon [sic], from being accustomed to our manners, and understanding a little English, was the person through whom we wished to prosecute inquiry; [. . .] Farther to please him, a brick house, of 12 feet square, was built for his use, and for that of such of his countrymen as might chuse to reside in it, [. . .]
>
> (Tench 1793: 200)

Clearly, something was wanted in return. It is not recorded whether Bennelong asked for the land – his request was just for a house – he surely presumed the land was his anyway and not Phillip's to give. In 1962, historian Marjorie Barnard summed up, dispassionately but concisely:

> The history of land tenure in Australia began very simply with one assumption: all land was the property of the Crown. The aborigines alone might have been supposed to have some claim upon it and they, probably, at that time no more thought of a right to the earth they trod than to the air they breathed.
>
> (Barnard 1962: 268)

The Sydney Opera House

Now, just as the house has disappeared, much of the long rocky promontory where it sat has gone as well, replaced successively by a number of buildings. The current occupant, the Sydney Opera House, has completed the process. This building's substitution of itself for the Point is total – no trace remains, there is nothing that can be seen that might now be called 'the site' of Bennelong's house; there is only building. And just as the site itself has gone, the name Bennelong Point may well be slipping from view as well; it remains on official maps but on many electronic maps, ubiquitous now, the Point is named only as the Sydney Opera House, the address no longer Bennelong Point but Macquarie Street.[3]

Despite this, over the years since its commissioning, links between the building and an Aboriginal past have often been invoked to secure and cement a connection between the Opera House and national identity. At the ceremony in 1958 to mark the beginning of construction, the wife of the Danish Consul-General to Australia presented the building's architect, Jørn Utzon, with a didgeridoo. In this gesture, albeit a small one by a very minor player, it is as if the didgeridoo, baton-like, was being used to pass the Point on to Utzon; the place associated for so long with Bennelong given over to someone else, as if in a relay.

Further, the official souvenir of the opening on 20 October 1973 noted that Bennelong Point was the site:

> [. . .] of the first building constructed for aborigines by Europeans in Australia. It is named after Bennelong, a notable figure in Australia's early history with whom the founders of the country first made contact in 1790. Bennelong maintained a long and intimate association with the early settlers and the brick hut built by Governor Phillip became a recognised gathering place for aborigines at Sydney Cove. There the colonists saw a corroboree for the first time and Bennelong was one of the masters of ceremonies.
>
> (Sydney Opera House Trust 1973: unpaginated)

A picture arises from this of a place ordered and shaped by the Europeans and one of the ordering devices is the building of a house. There is a sense of Bennelong innocently encountered rather than kidnapped, of the cultivating and sustaining of relationships without any hint of complexity, pressure or exchange. And the house – built by Phillip – in this telling, is not a model home, it has, if anything, the flavour of a community centre, a service provided by the government for what are now, through no agency of their own, its subjects.

The passage above also establishes the Point as a place where, historically, performances are conducted and so the Sydney Opera House seamlessly continues an ongoing tradition. In 2006 the Opera House website suggested a seamless transition between the two cultures, starting with Bennelong's return to Sydney Cove after the spearing of Phillip:

> He later cautiously re-established contact with Phillip testing the Governor's intentions with numerous requests for favours, including a house to be built on the end of Bennelong Point.
>
> In 1790 Bennelong took possession of the hut ordered for him by the Governor. The hut was used as a social centre by those Aborigines that frequented the settlement.
>
> The first known concert on Bennelong Point was in March 1791. Bennelong provided an evening of entertainment at his house for the Governor and his party.
>
> (Sydney Opera House Trust 2006)

Today, the website says less: 'In 1791 [sic], a hut for Bennelong is built on the site where Sydney Opera House now stands' (Sydney Opera House 2012). But, implicit in such phrases as the earlier 'requests for favours' and 'ordered for him by the Governor', even 'a hut is built on the site', is an idea of largesse that conceals the process of giving with one hand while taking with the other. Here, the heritage of the Opera House is linked back to the house not as the house requested by Bennelong, but as the house built for Bennelong.

These casual sentences and the attendant suggestion of generosity obscure what Bennelong and his compatriots were being made, covertly, to accede. A certain degree of status and, by extension, ownership, is accorded to the builder, and thus the idea that the house was enmeshed in a web of inducements and coercions, both economic and moral, drifts away.

Bennelong's house takes on something of the aura then of what anthropologist Annette Weiner terms an 'inalienable object' (Weiner 1992). That is, those things whose value is primarily determined by their provenance, like old watches or cedar chairs, Captain Cook's cup and saucer or a petticoat once worn by Queen Victoria. It is unusual perhaps to consider that 'inalienability' could be preserved in an object no longer in existence, but Bennelong's house, I maintain, has been kept alive with his name and by the continued circulation of the story of the house. This gives it power to enact the primary value of inalienability, which is

> [. . .] expressed through the power these objects have to define who one is in a historical sense. The object acts as a vehicle for bringing past time into the present, so that the histories of ancestors, titles, or mythological events become an intimate part of a person's present identity.
>
> (Weiner 1985: 210)

There is no 'person' here whose identity is being constructed though, it is rather the Sydney Opera House in whose name the house's inalienable status is defined historically.

The Sydney Opera House is a monumental building and just as Phillip saw Bennelong's house as a step in the direction of civilizing the Aboriginals, so the Opera House presents to the world the extent to which Australia has taken on the mantle of European culture. Symbolized within it is the power of will to shape place, and the massive reworking of Bennelong Point could be seen as being analogous to the way the whole country has been reshaped and re-inscribed. And, just as the British brought their own customs and culture with them to this place, so the Opera House's architect brought his. Despite the site's history there is little in the Opera House that references historic aspects of the site, or that suggests any emotional resonances those may have left, and the Opera House design competition brief made no requirement that the site's history should be acknowledged in any way. If the brief suggested a monumental edifice was appropriate, it was because the setting was monumental, as were the nation's aspirations.

The building is a wondrous response to its physical site – to the harbour, the harbour bridge and to the city, but its physical presence has been shaped by an assorted collection of international architectural references. It is only the 'shell' likeness of the roof design that might point, unintentionally, to the shell middens the British noted in the area. But while the idea of shells or sailing boat sails might visually link the building to its setting, almost all the shape and material references are from somewhere else. The journalist Gavin Souter, reporting Utzon's first visit to the site, noted: '"It's right," said Utzon in his lilting Danish accent, looking up the Harbour towards Pinchgut. "It's OK. This is the way they placed the temples in the old days"' (Souter 1973: 10).

The podium owes much to Mayan and Asian temple architecture, and these inspired Utzon to conceptualize an elevated platform as a place where the concert-goer was separated from the ordinariness of everyday life. It was to mark the place apart from its earthy surrounds and prepare the user for immersion in the imaginative realm that theatre, concert or opera promises. The podium is encased in aggregated granite slabs, whereas if memory of place inheres in any material in Sydney, that material is sandstone. Working from photos, the architect found the Harbour to be dark, and the white of the roof would catch the sun as it did, he said, in 'the Mediterranean or South America or other sunlit countries' (Messent 1997: 104).

Weston notes that Utzon hoped the stoneware roof tiles would give Danish visitors (though the tiles were made in Sweden) a feeling of being part of the building (Weston 2002: 120). Particular clays can represent traits or aesthetics valued by specific cultures; porcelain clay is typically linked to the East, but stoneware has long been associated with Germany and Scandinavian countries where it is much used. As a material it is marked by its toughness, its rock-like qualities, and its rightness for specific functional tasks.

So the tiles, through their true stoneware-like qualities, establish a relationship to Danishness that is added to the myriad cultural references. So too is Kronberg castle near Utzon's home at Helsingor (Hamlet's Elsinor) cited as being an 'analogue for the site in Sydney' (Weston 2002: 120). To the design references above can be added Utzon's interest in systems of growth and the more concrete mathematical studies he made of the surroundings. Phillip Drew notes that prior to making the design Utzon 'began by studying sea charts of Sydney Harbour, from these he measured distances to assess the height of the surroundings to develop a feel for the landscape' (Drew 2002: 14).

In balancing the natural features of the harbour, Utzon appears to have worked in the manner of a painter or sculptor bent on finding just the right solution to complement or improve upon the composition the natural topography offered.

Walking through the Opera House one is struck by the way in which, as you move around the outside walks, climb the podium and travel through the interior, vistas open up, views are framed, and the drama of the Harbour is offered up. The various outlooks coupled with the intrigue and pleasures of the building itself call you to map the space on foot, and, as you are led along and to its views it can seem that your progress is being orchestrated, in a 'Capability Brown' sort of way, through a series of picturesque events. And that it should be a gardening experience associated with eighteenth century Britain that comes to mind is a reminder that the identity Bennelong's house gave to the Point so long ago was also British.

Styling Bennelong's small house as a 'model home' as I did earlier resonates with the Sydney Opera House. The earlier house, in colonial thinking, might have marked the place where Bennelong and his compatriots were rescued from a remote, uncivilized past. The Sydney Opera House too might model a more cultured Australia, distanced from its uncivilized (convict) past. The identity fashioned here is one also firmly attached to a European past.

It is impossible to imagine how life must have been for Bennelong and his countrymen and women at Sydney Cove after January 1788, but it could be imagined that the first section of the title given to Keith Willey's book (Willey 1979) about the clash between the two cultures, *When the Sky Fell Down*, might describe it. In the larger scheme of things, the circumstance of Bennelong's house might seem minor, and the magnificence of the Sydney Opera House is not in question here. Nevertheless, a type of complaisant disregard accumulates each time words such as 'a house is built for Bennelong' are aired for it was one of the many small acts that presaged, to use now the second section of Willey's title, *The Destruction of the Tribes of the Sydney Region, 1788–1850's* (Willey 1979).

At the 2004 Biennale of Sydney, the American artist Jimmie Durham presented a performance on the forecourt of the Sydney Opera House involving a large marble boulder – onto which a face with a somewhat quizzical expression had been painted – held aloft above a small, bright red and sparklingly new sports car. The performance consisted of the painstakingly slow lowering of the marble missile onto the car until it buckled under the rock's weight. With its roof caved in and its chassis resting on the ground, the car sat there for the duration of the Biennale with its grotesquely obese passenger, the rock, appearing bemused by the devastation wrought by its weight.

In the Biennale catalogue, Michael Taussig connects Durham's stone to the stone used in an Aboriginal initiation ceremony performed close to where the Opera House stands and documented in 1790 by Lieutenant-Governor Collins:

> [. . .] the young men were hoisted on the shoulders of the men to have a tooth knocked out . . . To complete the extraction, a wooden chisel hammered by a large stone was used, the man making a 'most hideous noise' as blood was allowed to freely run down over the head and shoulders of the man carrying the young man who would henceforth bear as part of his name, the name of the man carrying him.
>
> (Taussig 2004: 82)

The most monumental building – with its own compelling narrative – now sits on the shoulders of Bennelong's house and symbolized within it is the power of will to shape place. Bennelong's hold on the Point might be slipping from view and the public historical sphere might weave the narrative of the house to suit any number of purposes at will. But that small brick house, securely cemented in the eyewitness accounts of the time, speaks of something undeniable.

Interviewed at the time his work was performed at the Opera House, Jimmie Durham spoke poetically about his ideas on the complex interactions between objects and words:

> I had [Jose Saramago's] book about the death of Ricardo Reis and for a month I walked around Lisbon, my first time in Lisbon and I gathered objects and quotations from his beautiful book on to these things. So there's a construction made of some metal and some abalone shell and some wood, and it says, 'Do you accuse me of lying? No, we never lie to each other we are friends. When we need to lie we use words, they lie for us.' And this little construction of things I found began to do that in such a nice way. They're not lying, but they're calling it to attention that something is lying.
>
> (Durham 2004: 49)

Do buildings lie? Can they be articulate? Perhaps, rather, they are enabled to perform through the interdependence of buildings and texts; not mute, they are able to narrate a multiplicity of histories and identities. The overlaying of one civilization on top of another in this country in the closing days of the eighteenth century had profound implications for the original inhabitants and so the questioning of the primarily British foundational narratives remains an ongoing process.

In twenty-first-century Australia, as in most places, issues surrounding national and private identity are complex and difficult to unravel, but what I have attempted to show in this chapter is that it is at least possible to step back in time and see – with some clarity – the ways in which identity was differently performed over time by a small brick house.

Chapter 3

Emplacement, embodiment and ritual

Some considerations from *shikii wo matagu* for our understanding of place and identity

Robert Brown

Introduction

Is architecture as important as we like to think it is? While this proposition intrinsically challenges the emphasis attributed to architecture in framing our lives, the intention is not to negate the role architecture (in a conventional formal sense) can and does play. Rather, the aim is to explore the role of other, everyday factors and prompt deliberation of a more heterogeneous understanding of place-making.

To pursue this enquiry, one particular example of spatial representation has been adopted as a point of departure. *Shikii wo matagu* is a ritualized act of movement across the threshold within Japanese culture. Despite significant changes to the physical environment, and in *what* environment Japanese are living, *shikii wo matagu* maintains its relevance in contemporary Japanese life. The nature of this act, and its underlying concepts, values and social relationships will be explored within this text. The implications of this for our understanding of place and identity will equally be considered.

A seemingly simple act of entry

Shikii wo matugu describes a seemingly simple undertaking of crossing (*matagu*) the threshold (*shikii*) (Digital Daijisen 2011a). Though related to entry to and departure from the home, it is not limited to the home, and applies to other conditions, from the sanctified space of a teahouse to quotidian settings, such as a restaurant. Moreover, the phrase is applicable to movement across any boundary between 'inside' and 'outside', though the home is the main point of reference here.

While *shikii wo matagu* is a phrase which is not expressed in everyday conversation, its potency is underpinned through the exchange of commonplace yet distinct salutations offered as one crosses the threshold. This language is paralleled though the body's engagement in ordinary, though again distinct, gestures, notably removing one's shoes before entering the home. Other gestures may also form part of the sequence of movements associated with *shikii wo matagu*, including: approaching a screen (or series of screens)[1] in the external skin; opening the screen(s); stepping into an entry chamber (*genkan*) (Digital Daijisen 2011b); turning to close the screen(s); sitting on the edge of the raised internal floor of the dwelling (*agarigamachi*) and removing one's shoes; and/or standing back up and stepping up onto the raised internal floor (Digital Daijisen 2011c).

Also associated with *Shikii wo matagu* is *agate kudasai*;[2] literally meaning 'step up', it relates to the act of stepping up onto the raised floor of the interior. While relevant to *shikii wo matagu*, *agate kudasai* is often used as an expression of invitation or welcome into one's home. Linguistically, it evokes the use of the vertical to distinguish spatial hierarchy and with it social standing, and to delineate the sacred from the profane.

The origin of *shikii wo matugu* is not clear; it may lie in the substantial raised entry threshold in early rural dwellings. This threshold most likely served a practical purpose of keeping dirt out before later developments such as the change

Figure 3.1 Associat with *shikii wo matag* is *agate kudasai,* literally meaning 'ste up'; this action is paralleled though th body's engagement ordinary, though dis gestures, notably removing one's sho before entering the home.

in level appeared (Cornell 1997: 25–27). Over time this level change, together with other physical demarcations outlined above, became further articulated. By the late 1800s these coalesced to give the form of entry commonly associated with the *traditional*[3] Japanese house.

In contemporary dwellings, however, these are mostly absent. Various factors have contributed to their disappearance, notably the need for level (i.e., mobility-impaired) access. Further impetus for the entry's physical transformation has been as a corollary to changes to the once multi-layered external envelope of *traditional* buildings, arising from: spatial demands in dense urban settings; new material capacities; and changing expectations of light and thermal requirements (Brown 2008: 7). For obvious reasons, the *traditional* entry setting is not found in the homes of the increased numbers of Japanese living overseas.

Despite the loss of these physical markers, *shikii wo matagu's* relevance endures. This ongoing presence is, however, not achieved through contemporary reinterpretations of *traditional* forms, but rather through how Japanese people conceptualize and inhabit their environment. While sometimes supported through the demarcation of a threshold by placing a mat on the floor, the positioning of furniture or the appropriation of a joint in the floor finish, even where no physical line exists a transitional space still operates in which *shikii wo matagu* is enacted. This suggests that the act and its meaning are not singularly dependent upon any physical signification, and are also grounded in other representations.

Ritual, emplacement and embodiment

Belying the seemingly simple gesture its name suggests, through its practice underlying beliefs and values are brought into action and exposed, a revealing indication of a process of ritualization (Kawano 2005: 4). We understand ritualization as a way of acting to distinguish one act from another (Bell 1992: 74); concurrently, ritualization sets its site of enactment apart from other locations.

What is of most significance here, however, are not any physical movements involved in the ritual itself; as everyday gestures they do not attract a significant amount of attention (unless one does not take one's shoes off at the entry to a Japanese home). What is distinctive is what the ritual does, both in terms of its meaning as a socio-cultural and spatial representation, and how this meaning is enabled.

Shikii wo matagu generates and reinforces a conceptualization of inside–outside within Japanese culture, and so distinguishes a sense of place and fosters a sense of identity and belonging. A distinction between inside and outside figures strongly in Japanese society, assuming a pre-eminent role in structuring their world.[4] While associated with dirt, germs and pollution, in a broader sense outside (*soto*) is conceptualized as a place of crowds, strangers and anonymity, while inside (*uchi*) is synonymous with one's household or those within one's group (e.g., relatives, classmates, work colleagues).[5] This meaning is engendered early in childhood (Hendry 2003: 47), with *uchi* associated with security, support and belonging, while *soto* is connected with the converse.

This inside–outside dialectic is further manifested through social relationships in the concepts of *honne*, one's private feelings or 'face', and *tatemae*, one's public 'face'. On each side of the boundary, distinction is made regarding appropriate behaviour and responsibilities towards one's family or group, and towards those outside it (Hendry 2003: 49–50). One example is in patterns of speech, expressed via degrees of informality/formality, and in the words and actions employed to communicate social obligations towards others. It is further evidenced in what is considered appropriate to reveal in conversation, from one's thoughts to one's emotions.

This articulation extends to the exchange of greetings and goodbyes as one crosses the threshold.[6] These phases imply underlying attitudes about taking care of dangers while outside and welcoming one back from those dangers upon returning. Even for those outside of the family or group, expressions of goodbye communicate a sense of taking care as they journey back home or onward. This language's role should not be underestimated, and is significant in two respects: firstly, language transforms and mediates body-focused actions into conscious awareness and the experience of a particular subject; secondly, it enables this awareness and experience to be communicated to others and so establish socio-cultural norms (Noland 2009).

What *shikii wo matagu* also does as a ritualized act is to spatially situate this meaning. This is realized in a number of ways, involving the ritual's site, the body and an explicit manipulation of both (whether consciously or unconsciously); these occur through what are termed emplacement, embodiment, and transmutable agency, and are further explored below.

Through ritualization meaning becomes emplaced in a setting as one comes to associate, know and remember it as the site of this ritual (Kawano 2005). In so doing, the participant conceptually re-schematizes the environment, identifying it with the ritualized act (Bell 1992). This structuring (i.e., production) of space allows us to orient and situate ourselves within it (Leach 2005). It does not impose a totalizing structure, but it does allow us to make sense of and operate within the world (Bourdieu 1977: 123–124). Key here is the act of emplacement; while the nature of the space may/will inform how we interpret it and operate within it, as active participants we are not reduced to merely responding deterministically to what the space objectively gives to us. In contrast to some interpretations of affordance (*cf.* Gibson 1977; Gibson 1979), the meaning of the space is not solely latent within it and realized simply through our reaction to it; rather we subjectively emplace upon the environment our own understanding built up through prior experience, beliefs and values.

Most of the time we are not aware of this emplacement as part of daily practice; this posing of schema is, however, often marked through ritual practice and, through this, awareness of personal and/or communal values may be delineated (Tuan 1991: 36–37). Bourdieu's construct of *habitus* reflects this; through ritual, agents master, i.e., reaffirm and/or gain understanding of, their *habitus* (Bourdieu 1977). It is important to note that these cognitive structures are not innate, but rather are acquired, notably in childhood, and evolve

re 3.2 Through ·lization meaning omes emplaced in tting as one comes ssociate, know and ember it as the site ·is ritual.

through experience and conscious as well as unconscious thought (Bourdieu 2005: 45).

Simultaneous with this emplacement is that the ritual becomes rooted in the body. The various movements involved explicitly engage the body, prompting the participant to be more aware of their own body within the setting. The body's involvement accentuates the division between inside and outside, and is further embedded through the normalizing of the act as a code of behaviour. These actions produce a ritualized body 'invested with a sense of ritual' (Hendry 2003: 98). These norms are further extended outside the body, such as through not bringing into the interior shoes soiled by the outside. Through this grounding, the underlying concepts and values are not independent of the body but are themselves embodied, i.e., the body is not separate from the structuring of space as some other object acting within it, but is instead a constituent part of the conception of space. The body is thus a site for communicating complex thoughts and values that are represented through the body (Farnell 1995: 18, 23, cited in Kawano 2005: 7). We can understand the body not as a passive vehicle for cultural experience, but rather as an active participant within it. It is an entity capable of experiencing and synthesizing that experience, and not simply a container for the mind as the primary site of experience and understanding (Merleau-Ponty 2002). The body is thus not only culturally shaped but also 'supports yet inflects culturally legible signification', while culture equally is 'both embodied and challenged through corporeal performance' (Noland 2009: 2, 20).

Though the above is framed within a Western perspective of the body, within Japanese culture, the body's role as a site of cultural expression has both historical and continuing significance.[7] Discourse on Japan has positioned the body as central to how the Japanese have and continue to engage with the world; this contrasts with Western philosophy (at least since Descartes), which has tended to privilege the mind, with the body as just another representation of the outside (Parkes 1995: 81; Taylor 1999: 32). Within Japanese *tradition* however, the body is not understood as something that 'distorts the judgement of the intellect but as a fundamental and trustworthy part of what it means to be human' (Tanaka and Tanaka 1998: 187). Central to this awareness is a sense of the body as a site for disciplining not just the body, but equally one's inner self, and through this both cultivating both the whole self and producing a moral self (Kawano 2005: 39–40; Parkes 1995: 93). Discussion of such body-based moral production is typically associated with practices such as seemingly overtly physical practices such as the martial arts, yet equally with more ephemeral performances such as the tea ceremony (Parkes 1995: 82). It is also present in other forms of body practice, most notably here that of shared quotidian ritual praxis. Through such rituals, communal values find expression in acting bodies (Kawano 2005: 39); concurrently an acting body comes to embody those communal values (Kawano 2005: 7). Reflecting Japanese philosophical thought, shared and embodied sensibilities are neither rendered as abstract conceptualizations nor somehow separate, 'in isolation from life, but [. . .] rather [are] embodied in particular forms of practice' (Parkes 1995: 79). As Reader posits, 'Japanese people today are known to emphasize "the primacy of action" over belief in explaining their ritual actions' (Reader 1991: 15, 20, cited in Kawano 2005: 1). As Kawano further articulates, body-based practice brings 'ideas and values into action, thereby (re)producing them (Kawano 2005: 4).

The emplacement of ritualized meaning, and the simultaneous embodiment of meaning, inform each other with what Bell cites as critical circularity, i.e., the body's actions and embodied meaning redefine space, while simultaneously through generating this space, the body's movements are structured (Bell 1992: 93). Noland similarly discusses this as 'distributed agency' (Noland 2009: 110), while Nast and Pile note that 'bodies and places are corpo*realized* through the embodied spatial practices' (Nast and Pile 1998: 408). These constructs are echoed in further discourse, with the body posited as both constituted by and as a constituting object in relation to other objects (Merleau-Ponty 2002: 237); Bourdieu meanwhile finds in the dialectical body–space relationship structured by ritual an exemplar of embodying structures of the world, 'that is, the appropriating by the world of a body thus enabled to appropriate the world' (Bourdieu 1977: 89).

Intrinsic to ritualization, and concurrently to the constructs of emplacement and embodiment, is a capacity for change. Central to this is that the schema engendered through ritualized acts is not just limited to the site of the ritual; rather, they can generate and support a worldview that structures social relationships and engagement with everyday existence beyond the site. Ritual infuses a sense of order into daily life, and enables people to 'fashion a strategic context for interaction and appropriation' (Kawano 2005: 8). Bridge finds an element of defensive structuring in this, in that ritual provides a safe, strategic place from which to act on a range of conditions in a wider context (Bridge 2005). Moreover, ritualized practices, once internalized, are capable of being strategically manipulated and can respond to new conditions of place, time, technology and socio-cultural relationships (Bell 1992: 107).

Such arguments indicate that ritual practice, and practices of cognitive spatial structuring and body performance, are not fixed but rather are more malleable and have a capacity for transmutable agency, i.e., they have a sense of purpose and are capable of (re)structuring place and experience beyond the specific (i.e., original) site and body gestures. While informed by past conditions and experiences, they enable us to cope with changing situations in the future (Bourdieu 1977: 83); equally, we can adapt them to other contexts (Jenkins 2002: 78). Indeed, they are best understood as not only transformable but creative and generative (Bell 1992: 92; Hillier and Rooksby 2005: 11; Noland 2009: 74).

Displacement and decorporealization

To consider better both the significance of *shikii wo matagu* and the implications it poses for our understanding of place and identity, it is relevant to situate it in the context of the contemporary Japanese city; this landscape in many ways exemplifies the construct of a modern, globalized world, one which would seemingly threaten, if not negate, the ongoing validity of *shikii wo matagu* and its meaning. Echoing discourse on modernity and its various mutations, electronic images and text as graphic image assault one's perspective (Auge 1995; Armitage 2000). The city's spatial-form reflects discussion of a shift towards an emphasis on the processing of information over experience (Heynen 1999: 156); it is equally suggestive of McQuire's critique of Virilio on the electronic image's undermining of place (McQuire 2000). Such a framing echoes the proposition that 'architecture can function in a similar way to an installation, giving priority to the display of instantaneously and visually absorbed information' (Toyo Ito, cited in Knabe and Noenning 1999: 97).

Within this landscape the body has become lost, if not negated, what Bridge has defined as the city being 'de-corporealized' (Bridge 2005: 15). This loss of the body as the measure of space arguably extends back to the advent of a Cartesian sense of space (Lefebvre 1991: 1). It is, however, in the contemporary (Japanese) urban landscape that we see the fullest representation to date of the de-corporealized city, in which haptic experience has been displaced by '. . . a complementary value: scintillating symbolic existence' (Shinohara 1982: 115, cited in Breslin 1988: 27). It is further mirrored in the view that 'three-dimensional existence is dissolving into abstract, image-life surfaces' (Okabe 2000: 4). This distancing of the inhabitant from their surroundings echoes references to 'the gaze the modern subject directs on his environment,' an objectified world from which he is separated and has no direct access (Berque 1997: 64).[8]

The physicality of the urban landscape is itself distorted through a dialectic of simultaneous fragmentation and homogenization. Reflecting changing social mores and fueled by massive and hyper-paced development, most particularly during the years of the economic miracle and bubble-economy but still continuing today, the sprawling urban landscape devolved 'not as a rationally programmable artifice, but as an increasingly irrational, often hostile, environment', in which design has become 'limited to individual buildings, as "show pieces", with a metaphorically and metonymically transformed and often fragmented imagery' (Yatsuka 1988: 7). Architects working in this chaotic condition rationalize that the only viable response is '. . . to add to the restless image of the city' (Bognar 1990: 14). Simultaneously, a universalized sameness is projected by the commodified nature of the landscape of imported and home-grown franchise stores and restaurants, whose architecture (as symbol and/or sign) is reproducible at any time and anywhere. Within such contexts, Tuan argues that not only is spatial meaning diluted but it does not even intend to present a world-view (Tuan 1991: 112). This observation reverberates in Bognar's commentary on contemporary Japanese architects pursuing a reactionary 'critical path of practice that *destructures* Form and Meaning' (Bognar 1988: 20).

The sense of displacement in the physical landscape has been paralleled by changes to the socio-cultural fabric. Notions of place and identity were once tied strongly to the community in which one lived; via the mobility of modern transportation, most Japanese now live in urban areas in which interaction with other people is based more on the people they meet through work or recreation than on the immediate neighbourhood in which they live (Hendry 2003: 74). This has further accelerated via the availability of mass media, telecommunications and imported goods and ideas. Definitions of place have shifted, and with it inherited spatial constructs of cultural identity. We now live in a 'global supermarket', in which our identity may be linked more to others across the globe who share our beliefs, interests and tastes than tied to any particular region or culture (Mathews 2000). Concurrent with this has been a sizeable shift in demographics, with the number of foreigners residing in Japan having more than doubled in the last 20 years, 1989–2009, while the number of Japanese living overseas has almost doubled as well (Statistics Bureau 2011).

Further discourse contends that these forces' impact is so pervasive that fundamental constructs of identity and place have become questioned, diffused or even negated. We are so dislocated from any normative sense of place, it is argued, that 'individuals may have difficulty in locating a sense of belonging' (Kim Dovey, cited in Roderick 1995: 61). A further challenge contests that even the dialectic of inside and outside is no longer relevant (Hardt and Negri 2000: 217). Jameson carries this forward, noting our spatial environments have simultaneously overwhelmed and undermined 'the capacities of the individual human body to locate itself, to organize its immediate surroundings perceptually, and cognitively to map its position in a mappable external world' (Jameson 1991: 44).

In parallel with these thoughts are questions about the presence of *tradition*[9] within contemporary Japanese life. There is a (Western) tendency of defining Japan through its past, and failing to fully recognize what contemporary Japan has become. This is particularly evidenced in architecture, where Western interpretations of *traditional* architecture often frame discussion of contemporary buildings; yet as architect Kazuyo Seijima relates, 'it is Westerners who analyze Japanese architecture in these terms more than we do' (Jodido 1997: 72). This non-*traditional* orientation is further illustrated in native aspirations towards an 'emphasis not on any "unique Japanese-ness" but rather on ideas that obtain universality' (Taki 1988: 32). While *traditional* culture is still present within the landscape, the Japanese have become like urban dwellers anywhere with tenuous everyday connections to any (mythologized) past (Martinez 2005).

Implications for our understanding of place and identity

In spite of numerous changes, from socio-cultural trends operating within and outside Japan, to dramatic alterations to the spatial-form of the urban landscape and domestic interiors, *shikii wo matagu* has maintained its relevance. This is testimony to the potency of emplacement and embodiment (and ritual practices within which they are embedded); its resilience is also evidence of its capacity for transmutable agency. Through repetition of the act and a resultant appropriation of space, Japanese people living both in Japan and overseas continue to project a culturally embedded cognitive spatial structure and body awareness onto the world, and so define their collective identity within it. This suggests that our constructs of place, notably of inside and outside, retain validity, and that there is still legitimacy in the value we place on what Tuan defines as a sense of 'rootedness' (Tuan 1991: 153), and what Lovell refers to as a sense of 'belonging' (Lovell 1998).

e 3.3 Our
tructs of place,
ly of inside and
de, still retain
macy, and reinforce
se of belonging
ootedness;
wa House, Tezuka
itects.

The central point however is not to adulate *shikii wo matagu*, nor to deny that powerful universalizing force which might displace us from our conventional notions of place and identity are acting upon us. Indeed, if anything, I would posit Auge's argument that these opposing tangents, which would seemingly negate the other, are in fact palimpsests that overlap, touch, reflect and/or synthesize each other (Auge 1995: 79). Rather than one narrative replacing the other, they engender a collagist construction of place (Giddens 1990: 108, cited in Dovey 2008: 49). In this context of multiplicity, what lessons might we draw from *shikii wo matagu* about our understanding of the making of place?

First, we are able to generate a sense of place (and sense of identity) not only through physical production, but equally through emplacing a structuring, cognitive representation onto space. Central to our inhabitation of place is a proactive process of making (and remaking) meaning in the everyday (Heynen 1999). In response to the designed environment as a pre-existing condition, people can be adaptive, determined not only by universal, abstract, ideological forces, but also by their own concepts and actions. People are not simply part of some larger imposed conceptual body, and through everyday life turn abstract space into place (Bhabha 1990: 294). The spatial tactics shown by users are testimony to an understanding that place is not fixed or determinate, but rather that it can be a site of agency in which ideas can both be contested and constructed (Tonkiss 2005: 7). Such a malleable spatial conception is reflected in Japanese thought, in that 'space is not approached as a tangible act of a real existence' (Stewart 1987: 10); rather, as exemplified by the Shinto Shrine at Ise, spatial-form 'exists only as a phenomenon, that is, in the act of reconstruction' (Kuma 2000: 58).

Second, a critical aspect of this everyday making is located in the body. Various discourses attest to the centrality of the body in framing our sense of place and our identity, including as an anchorage in the world and as an expressive space itself (Merleau-Ponty 2002); a measuring point (Leach 2000: 79); a generator of dialogue with the world (Bloomer and Moore 1977); and/or as the locus for schematizing the world (Bridge 2005: 20). These roles, evidenced in our repeated actions and behaviours in the everyday, manifest what Butler calls the 'perfomative' (Butler 1999). Place and identity are thus not separate from the body, but are intrinsically constituted in it. Indeed, Casey argues that place is what occurs between the interaction of the body and the landscape (Casey 1993: 29, cited in Boeck 1998:46). Kawano further extends this argument, noting that bodies don't exist in a vacuum; rather body and place mutually constitute each other through the embodiment and emplacement of cultural beliefs and meaning (Kawano 2005: 7, 70).

Third, there is a need to recognize the potential of ritual for place-making. Ritual is accepted when looking at 'so-called primitive cultures', but is not acknowledged as an intrinsic part of modern secular society (Hester 1993: 279). Too often ritual is only acknowledged at the periphery of our understanding of architecture. It is even suggested that ritual has a reduced role in making place in modern life (Tuan 1991: 116). Yet ritual forms a vital part of experience in the contemporary every day. Hester's example from an otherwise unexceptional community demonstrates how a sense of place and community belonging is engendered through shared, ritualized practices in the public realm (Hester 1993). Butler similarly suggests a still potent authority is gained through the performance of ritual, maintaining that it remains one of the most influential practices by which 'subjects are formed and reformulated' (Butler 1999: 125). As Berque further attests, in Japan the experience of spatial meaning 'resides not so much in edifices (place) as in the perpetuation of acts' (Berque 1997: 94).

Fourth, this making in the everyday has the capacity to be transmutable, responding to new conditions and contexts. People bring their existing ways of seeing the world to any new site and are not limited to what the existing environment presents to them. In turn, not only might they not accept the given, but they may challenge it to pursue their own sense of agency. This poses a potential for emplacement and embodiment, and ritual, to serve as acts of resistance; in parallel, the places they engender can become sites of resistance (Nast and Pile 1998: 413; Noland 2009). In opposition to forces which might dislocate us, these acts can 'establish an emotionally safe relationship between men and their total environment by serving as organizers of activity and knowledge, as material for common memories which bind a group together, and as spatial referents for senses of familiarity' (Hart and Moore 1973: 274). Within Japan, such ritualized actions and the embodied meaning they emplace are subsequently found in other settings, linking 'ritual actors to other aspects of social life' (Kawano 2005: 115–116).

Conclusion

The intention of this paper has been to examine *shikii wo matagu* to pursue questions about the nature of place and identity. The former has revealed the ways in which this ritualized act contributes to how the Japanese structure the world, social relationships and themselves within it. This exploration has also raised thoughts about the nature of place and identity, i.e., that their making is framed not only through formal architecture, but equally through our own making; that this making is responsive to new conditions and challenges; that such making is grounded as equally in the everyday as it is in theorized ideologies; and that the body plays a central role in this making. Placing one's self in the world both cognitively and corporeally, structuring the world around ourselves and ourselves within it, remains central to the human condition.

We need to recognize the potential of people's capacity for this action. We contribute to the construction of our own narratives; this agency is scribed in place, and on our bodies, both literally and figuratively. The aim is not to negate the role of spatial-form in enabling, framing and enriching (or delimiting) our sense of belonging, but rather to raise the question of whether our writing of place and identity should be less fixed and more permeable. We need to understand architecture not only as what it is, but equally from the perspective of what it does.

Chapter 4

Architecture, architect and the recessive nature of technology

The work of Laurie Baker in India

Soumyen Bandyopadhyay

Introduction: identity, crisis, kitsch and technology

Arising essentially out of the colonial/imperial project – itself a product of liberal capitalism, either as a reaction to it or out of guilt conscience – the notion of 'identity' has been characterized by an anxiety regarding its imminent loss or disappearance. The development of post-mediaeval or modern 'high cultures' within colonized societies – such as the emergence of the Bengali urban culture in nineteenth-century Kolkata – continually subjected to the ebb and flow of colonial intellectual and cultural currents, was largely responsible for engendering this sense of loss and degeneration. As Marx had suggested, this had taken place due to a gradual distancing of the 'realm of culture (art, metaphysics, morality, ethics, law, etc.) [. . .] from the "real" conditions under which human beings produce their means of subsistence' (Abbinnett 2004: 74). The rift was significantly more than the critical distancing all cultures duly need for self re-evaluation and replenishment, providing the 'general determination of "culture as ideology"' (Abbinnett 2004: 74). The increasing distancing and abstraction produced an 'independent domain of human self-recognition' (Abbinnett 2004: 74), setting up a distanced and abstracted socio-cultural and political praxis as a window to look back at and exploit the original social context of cultural production. Enlightenment rationality, often propagated as instrumental thinking under the aegis of the colonial culture, empowered the emerging cultural centres to unilaterally 'de-culture' and ostracize its siblings and disown all 'natural ties' with its traditional roots. With the departure of the colonial patrons and the chasm that was by now unbridgeable through the technological empowerment of the urban elite, the identity crisis was complete.

While the anxiety regarding loss of legible identity has often given rise to conscious emphases on the visual expressive content, the architectural discipline has also had to contend with the development of reductivism and overt technological intentionality. The compromise has been a problematic attempt at masking such universalizing abstracting tendencies with the superficial expressions of identity, which in many cases fitted in well with the quick fix of identity politics. As Abbinnett quoting Adorno states:

> Ultimately, then, the secret tendency of culture to mask its complicity with the legal and political violence of its epoch is made manifest in the culture industry's reproduction of 'kitsch': those representations of the eternal forms of romance, heroism and sacrifice which draw us away from the barbarism of the present. Adorno put it in *Minima Moralia*: 'Today, when the consciousness of rulers is beginning to coincide with the overall tendency of society, the tension between culture and kitsch is breaking down [. . .] In administering the whole of mankind, it [kitsch] administers the breach between man and culture'.
>
> (Abbinnett 2004: 78)

This essay takes as its point of departure Frampton's suggestion that if architecture is to sustain itself as a critical practice it has to distance itself equally from the 'Enlightenment myth of progress and from a reactionary, unrealistic impulse to return to the architectonic forms of the preindustrial past' (Frampton 1985: 20). To break free from the shackles of the progress myth, Frampton urged architecture to position itself away from the blinkered pursuit of the 'optimization of advanced technology' (Frampton 1985: 20). This essay argues that the architecture of the British-born Indian architect, Laurie Baker, 'practicing' for much of his life from the city of Trivandram (Thiruvananthapuram) in the southern Indian state of Kerala, positions itself with a critical distance within this scenario. Baker, throughout his life, had searched for a legible and discernible localized identity in Indian architecture. This he achieved by, I would argue, inverting the above-mentioned architectural–cultural paradigm by closely aligning his architecture with the Indian social context of broader cultural production and allowing certain aspects of his work to be *recessive* or in a state of conscious withdrawal. Recessivity suggests not a complete absence but a degree of latency, a holding back; it is always there – *everywhere*, yet not really there – deflected further or made inaccessible. The recessive nature of Baker's work is evident on at least three levels: first, there was his virtually non-existent practice – in the institutional sense – which in effect enhanced reflective praxis by removing the ritualized insularity we have come to accept of late-modern architects and reducing his dependency on technological aid; secondly, his formal/spatial organization, which largely employed elementary forms and a consciously delimited vocabulary of articulation consisting of a few simple moves, positioned the work uneasily within Beaux Arts compositional expectations defined by taste and style; and finally, one could detect a recessive role of technology in his architecture in expressing active identity. This technological recessivity in his work, I argue, was closely aligned with his latent foreignness.

Recessivity and foregrounding

Leatherbarrow, in discussing the recessive nature of the horizon in *Uncommon Ground* (2002), gives us two reasons why the horizon is generally taken for granted in architectural thought: first that it is distant and therefore unlikely to be affected by design initiatives and secondly, it is an aspect of landscape that is difficult or even impossible to grasp. He refers to the similarly recessive nature of one's own shadow, in which 'all but its outline articulation remains hidden, everything else flattened, depth inaccessible' (Leatherbarrow 2002: 171). 'This sort of withdrawal', Leatherbarrow suggests:

> [. . .] is well known to anyone who has embarked from a seashore, for no matter how far one sails away from a port the horizon is always further ahead, seemingly soliciting but always outpacing further transit. For this reason it is senseless to see the horizon as a destination, because it can never be reached; [. . .] the horizon is always a potential site, its remote constancy reinforcing the sense one has of the radical contingency of human experience.
>
> (Leatherbarrow 2002: 172)

The notion of unattainability present in the ever-recessive nature of the horizon, I would argue, is at the heart of Baker's practice and architecture, continually breaking up and pushing back – deflecting – material, formal and tectonic expectations. It is presented to us in a state of the ordinary, and thus in unnoticed readiness, like serviceable equipment, as Letherbarrow tells us: 'when a setting works as it should, all of its parts coexist in a condition of shared latency or silent readiness' (Leatherbarrow 2002: 174). This echoes Hans-Georg Gadamer's understanding of our state of good health; we only realize we have a body when the body and its parts are overstretched during ill health (Gadamer 1996). Le Corbusier once observed that a proper servant is 'discreet and self-effacing, in order to "leave the master free"' (Leatherbarrow 2002: 174), and so does Kahn's understanding of servant and served spaces and components allude to this notion of readiness and latency.

With recessivity comes foregrounding; the receding horizon gradually reveals and foregrounds the intervening terrain. This is very appropriately conveyed in the well-known Heideggerian discussion on the evolution and nature of the word *building*, illustrating the manner in which language works; original meanings often recede into oblivion giving prominence to recent and foregrounded operative meanings. The Old English or the High German word for 'building' means 'to dwell', that is to remain or stay in place. The wider notion of dwelling becomes apparent when we realize that the root also means '[. . .] to cherish and protect, to preserve and care for, specifically to till the soil, to cultivate the vine' (Heidegger 1997a: 101). In this pervasive sense, dwelling simply means 'being on the earth', a meaning which remains embedded in everyday human actions and experience and therefore becomes a habitual act, giving rise to the ideas of habitation or inhabitation. Even the notion of the *institution* or establishment carries within it a dimension of *dwelling*. Thus, our awareness of the act of dwelling recedes behind the multitude of activities that constitute inhabitation on earth, reducing and distancing our conscious understanding of this central purpose in all our lives' actions.

Given that architectural sites are generally more precise targets than the illusive nature of horizon or language, the interplay between what is recessive or latent background and what is being foregrounded is more fluid and more intensely interwoven. The striking concentric, or the more distended curvilinear, organization of many of Baker's dwelling plans (e.g., Vaidyanathan 1972, Narayanan 1972–73, Paniker 1974, Alexander 1982; Figure 4.1) paradoxically foreground nature, either by allowing the natural realm to literally occupy centre stage framed by the architecture, or by adopting this form to respond to topographic constrictions. His elemental forms, square or circular in plan, result in tower houses of reduced footprint (Figure 4.2), which contain the architecture. Here, his architecture is material and mute, articulation and formal calisthenics kept to a minimum, distancing it as far as possible from the realms of language, and made meaningful only through foregrounded human inhabitation. Typological prejudices are cast aside in favour of providing a unique spatial organization suited to the client's needs and desires; yet latent archetypal spatial structures and meanings inherent in architectural elements are always detectable therein. This simultaneous phenomenon of receding and foregrounding characterizes creative actions and tectonic manifestations in Baker's architecture. In allowing his praxis – in its most expansive sense – to recede into a renewed relationship with the dense background of the everyday, he foregrounds the site, the life and aspirations of his clients and his architecture.

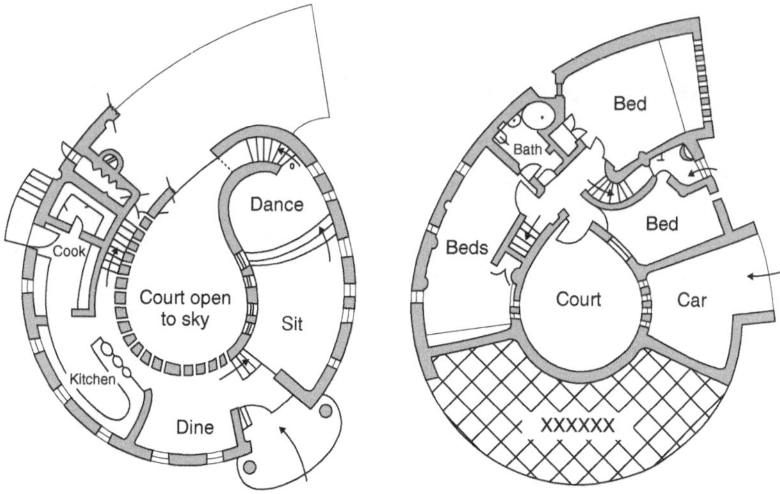

e 4.1 Narayanan
e, 1972–73; plans.

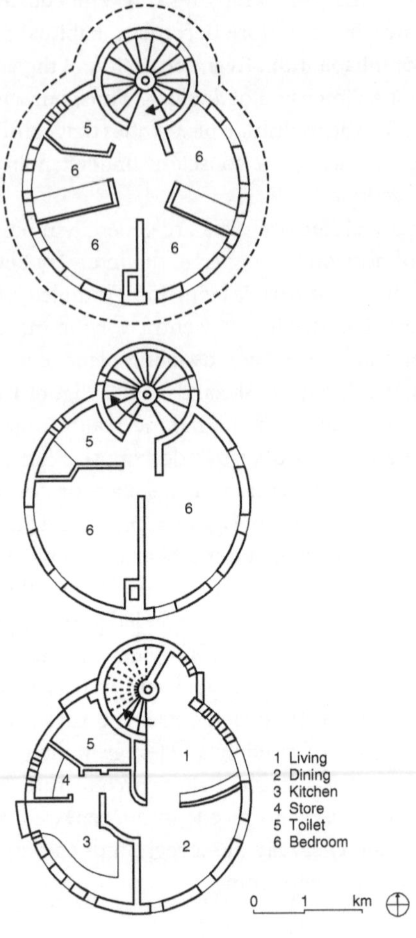

Figure 4.2 Namboodripad House, 1972–73, plans.

Legible identity and the problem of (over)abstraction

Dwelling, in the Heideggerian sense alluded to above, gives rise to our ability to simultaneously perceive the specific and the mutable on the one hand, and the ideal and the universal on the other, which, as Alberto Pérez Gómez suggests, is at the very basis of the creation of meaning in a world we inhabit. We are capable, as Pérez Gómez mentions, '[. . .] of perceiving the ideal in the specific' (Pérez Gómez 1988: 71).

The role of the arts, therefore, has always been concerned with the presencing of the ideal and the eternal, arising from the given and the specific, through the interpretation of our perception of a phenomenon; it is concerned with the *essence* and, therefore, with *abstraction* (Pérez Gómez 1988: 72). This was traditionally achieved through *mimesis* or imitation of the primary reality that surrounded us and subjected to perception. This reality, according to Husserl, could be described as 'intentional objects' that make up our world as a collection of givens, 'permanently open to both their individual specificity and their ideal essence' (Pérez Gómez 1988: 71). Analogy – geometrical, mathematical, linguistic, graphic – is an integral part of constructing relationships, the symbolic structure of which remains at the basis of *mimesis*, linking the ideal and the immediate world though our bodily perception. The perpetual state of becoming and approximation implicit in metaphoric analogy allows the reading of analogous representations being in a state of liminality, capable of being extended or recessed in either direction.

However, as Alberto Pérez Gómez indicates, 'technology developed as a utopia and became the universal mentality of the modern world, drawing its potency and pervasiveness from the human power of abstraction' (Pérez Gómez 1988: 73). Pérez Gómez questions Modernism's amorous affair with technology and instrumental thinking that emerged from mid-eighteenth-century theorizing – especially through Durand's work at the École Polytechnique – which indicated the beginnings of reductivism and overt technological intentionality. This reductivism – abstraction extended *ad infinitum* to nihilistic conclusions – has contributed to 'determinism and a calculated appropriation of reality', rather than the entire reality. Such intensely abstracted abstractions – of which Mondrian's paintings are good examples – have removed the 'primordial shared experiences of human beings as embodied, spatially oriented individuals', managing to achieve universality at the expense of meaning (Pérez Gómez 1988: 74).

In his writings and more explicitly through his architectural work, Baker had often wondered, 'what does "modern" mean?' (Bhatia 1994: 243). For Baker, the success of creativity lay not in the explication of the universal, rather in consciously ensuring its continued latency enshrined in the emergent 'intentional object', seeking its meaningful place in the world of reality. Thus in Baker's Modernity, legible cultural continuity was critical in ensuring the appropriate enshrining of the ideal dwelling his clients dreamt of. In foregrounding nature, in the manner we have seen earlier, Baker's work questions the instrumentality of technology and the theoretical disciplines within Modernism – and by extension, Modernist architecture's complicity in the effective domination of nature. Instrumentality, for him, resists technological latency, resulting in its superficiality, devoid of meaningful existence, as he highlights:

> [. . .] 'high technology' has [. . .] taught us that there is no need to concern ourselves with the weather or the functions for which the buildings will be used, or the variations in the cultural patterns of our clients – 'high technology' applique-work can cope with all this old-fashioned nonsense.
>
> (Bhatia 1994: 243)

Baker's (non-existent) practice

> [. . .] I have never run a proper architect's office. I have close to my bed a small, old drawing board – the same one I had in school. I broke my tee square quite a long time ago and never bought another. I have an old brass pair of compasses which belonged to my older brother and it was passed on to me when I first went to school. So I don't look the part at all!
>
> (Baker 2007: unpaginated)

Even though Baker had lived in India since 1945, he only set up something approximating a conventional practice in 1963 when he moved to Trivandrum (Bhatia 1994: 75), which, by his own admission, could be regarded as rather 'primitive'. The consciously rarefied nature of his architectural establishment and his minimal dependence on technology indicate a fundamental shift away from the image of the conventional modern architect we have grown accustomed to. For Baker:

> the most interesting part of designing a building is dealing with the clients – getting to know them, how they live and work and finding out what sort of a building they dream of. It is exciting to put on to paper what you think is in their heads, and then to go on altering or adding or deleting until you think you have put down what they want.
>
> (Baker 2007: unpaginated)

Drawings crystallized initial conceptions, which had to be developed and tested later on site. Thus, details were produced on the back of carefully cut envelopes, not with a view to those eventually appearing in attractive publications, but for the purposes of faithful translation from drawings to buildings (Baker 2007: unpaginated). This fragility and mobility of the graphic codification simultaneously acted in limiting his architectural practice establishment and extending it manifold. For, as the drawings lost their uniqueness and autonomy, the site assumed that role as much of the designing took place on site, involving Baker, the client and the builders. On numerous occasions he had commented on the wealth of knowledge he gained from traditional and local builders regarding the social appropriateness and cultural acceptability of local materials and building systems, forging a renewed partnership with an empowered traditional builder community. Reliance on local builders, his experimentation with innovative building techniques and his obsession with the desire to provide the unique dwelling or architecture for his clients at the lowest possible cost meant that mechanized building procurement was rendered obsolete. The site offered Baker insights he never had working at a drawing board as he once recounted, '[. . .] while I clamber about on the scaffolding, I suddenly realize that I will get a much better view, or more breeze, if I move the window or make it bigger. And so on' (Baker 2007: unpaginated). By limiting the role of technology and increasing public discourse, Baker's design process – his architectural *praxis* – rescued his architecture from technology's tendency to homogenize diversity, for as Pérez Gómez tells us, '[. . .] architectural *praxis* involves much more than technical means and scientific operations – it concerns values articulated through the stories that ground acts and deeds in a particular culture' (Pérez Gómez 2006: 205).

Baker's architecture and the recessive nature of technology

William Curtis, in discussing Mies van der Rohe's designs at the Illinois Institute of Technology (IIT) in Chicago, observed that:

> The local fire codes required that steel be coated in a layer of fireproofing, so that in order to express the structure 'honestly', the architect had to adopt the artifice of an extra veneer of steel around the fireproofing casing. At the corner of the buildings, this led to a curious detail in which the recessed core of steel was hinted at in a cut-away involving a steel veneer over concrete fireproofing, itself overlying the actual structure within the wall.
>
> (Curtis 1996: 401)

While Mies' approach to detailing has been variously praised for its structural clarity or Mondrainesque abstracted quality, it also reveals architecture's interesting position *vis à vis* technology. First, paradoxically, it achieves the expressive honesty through artifice, making the structure recede, allowing the expression of a selective set of characteristics, and then it reveals the inner steel core through the gaze of a cut-away section and the veil of a steel veneer, alluding to an ongoing dialogue between latent structure and the material appearance of the architectural enclosure. At a time when fashionable architects have virtually consigned architecture to a discussion of unitary and fragmented forms, seductive appearances and increased digitization of most aspects of its production and procurement, all underpinned by a particularly deterministic approach to technology, Mies' example reminds us of a key aspect of Baker's work and a fascinating window of opportunity almost on the brink of extinction.

The origin of the word *technology* lies in the Greek word *techne*, which they used both for craft and art. However, Heidegger reminds us that at its basis the Greek term denoted a 'mode of knowing' and not 'making'; a method of uncovering of beings. *Techne* as knowledge thus denotes a *bringing forth* of beings '*out of* concealedness and specifically into the unconcealedness of their appearance [. . .]' (Heidegger 1997b: 121), suggesting a strong parallel with the idea of presencing discussed earlier. Thus 'technology' simultaneously suggests stability and its ability to 'bring forth' or 'make architecture happen', indicating its fundamental or anchoring yet *recessive* or background role in bringing forth or constructing meaningful space and architecture.

For an architect who had devoted substantial attention to the development of low-cost building technology alternatives for *all*, is it not paradoxical to discuss technology's recessive nature within Baker's architecture? This, I would argue, has been made possible through an intense dialogue between enclosure and structure – a more densely interwoven condition than the one we evidenced at Mies' IIT. In the many variations of the low-cost enclosure he sought to reduce the cost of building walls significantly by making it a delicate and porous yet strengthened screen, plying and fusing the qualities of enclosure and structure into a single entity (Figure 4.3). Context, in Baker's architecture, following Deane's observation of Caruso St John's work

Figure 4.3 Centre for Development Studies, 1970–71, facade exter

> [. . .] ceases to be a passive tablecloth but a slow-moving inhuman agency at work within the design process. As a consequence the artefact absorbs and is even overwhelmed by site presences. It is as though the artefact were being returned to the edge of representation, to that primordial moment of figuration and becoming, when foreground and background are still interwoven or on the cusp of separation.
>
> (Deane 2009: 146)

This fusion is effected by light, that fundamental and primordial architectural ingredient, in turn enticing the brick enclosure to radiate its warmth and colour, expanding manifold its material aura (Figure 4.4). Porosity – and the resultant ambiguity between inside and outside – ushers in nature in more than one way. In this expanded and stretched materiality the structure is present yet latent; space is given a renewed density and gravity. The recycled, lightweight, terracotta filler tile inserts that reduce the cost of roof construction also add to this dense, thick atmosphere, where enclosure is fused and space moulded into a primordial chthonic plasticity.

Iconic and technological bias of modern architecture is cast aside, as component proliferation is restricted in favour of a simple yet intelligent monolithic integrity. Baker's architecture, thus conceived, contrary to the controlled procurement process of the modern architect exercised through a set of codified drawings proclaiming the power of superior tectonic knowledge, precipitated in a dialogue between technologies, within an expanded epistemological

ure 4.4 Centre for velopment Studies, 0–71, facade interior.

horizon. 'Finishing ends construction, weathering constructs finishes' (Mostafavi and Leatherbarrow 1993: 5). Baker's preference for an unrendered surface exposed his brick facades to an augmented material expression by inviting nature to invade its articulated porosity. Weathering – nature's way of partially reclaiming territory lost to human habitation – is part of a wider entropic process that all material existence is necessarily subjected to. As 'there is nothing necessarily pure or impure about dirt' (Mostafavi and Leatherbarrow 1993: 109), there is nothing inherently problematic about weathering, either. The process of foregrounding nature initiated by the spatial marks made on site is extended beyond the building's construction period as weathering takes over (Figure 4.5). By resisting 'finish', Baker perpetuated construction.

Epilogue: the 'foreigner' and detail as sites of dialogue

I would like to end this essay by reflecting on the agency of the 'foreigner'. Beyond the obvious technical value the 'foreign expert' adds to a project, the 'foreigner' is also capable of introducing a critical dimension of what I would call *a refreshed dialogue*, which is crucial in shaping active identity within the built environment. Much has been discussed about the generalizing properties of globalized interventions and application of expertise. Such interventions have been criticized mainly in the context of the neglect and erosion of the local and, evidently, universal application of highly specialized knowledge has resulted in the loss of cultural specificity. The built environment across the world has begun to look much the same.

Figure 4.5 Weatherin

Richard Sennett articulates a critical dilemma confronting modern man in a globalized society when he asserts:

> [. . .] the 'disenchantment of the world' [. . .] has meant that people experience the arrangement of society as mere conventions that can be changed at will. On the other hand, and perhaps because living in a truly disenchanted condition is more than most people could bear, modern society has generated a deep need to deny the ethical dimensions of disenchantment [. . .] Community, identity, roots: these human relations represent borders to be sealed rather than boundaries to be crossed.
>
> (Sennett 2002: 194)

Here Sennett contemplates the experiences and positions often held by the modern 'foreigner', the archetypal global citizen of today, which could be applied equally to the host culture. It seems that globalization has resulted in coexistent parallel worlds of host cultures and foreigners held at bay by invisible boundaries and divides. Misconceptions regarding the notion, role, opportunity and acceptance of the foreigner as an agent of interaction between cultures lie at the heart of the problem confronting cultures in today's fast globalizing world, of which the Middle East serves as a poignant example. The peculiar blasé disenchantment of today's foreigner is both a product of sensual and information overdose and a lingering aftertaste of colonialism. Sennett laments this denial of responsibility towards the subject culture, which has resulted from the cocooned and anaesthetized position of the foreigner. Today's globalized 'foreigner' is perched on the borders that Sennett describes – potentially opening up the possibility of dialogue between cultures and mediating and particularizing the globalizing influence.

All societies since time immemorial, as Sennett concedes, have feared and have felt threatened by the foreigner, but have also often apportioned them a privileged position; the mediaeval foreigner, on his part, like Boccaccio's voyeur in his *Decameron* stories, indulged in the passionate and self-enriching – perhaps even erotic – study of alien cultures and societies, bringing their own cultural position and beliefs to bear upon the subject. Colonial enterprise gave the colonizing foreigner controlling influence over other cultures and landscapes, empowering him with selective representation, romantic or otherwise.

The position of the foreigner within a society and the associated institutions has, at best, remained a subject of guarded curiosity and has always received circumspect treatment. The latter is also partly because the foreigner, as Georg Simmel believed, calls into question and exposes the arbitrariness of the 'society's script', held sacrosanct and preordained by the insider (Simmel in Sennett 2002: 191). Baker's existence in India could be termed as one of willing exile – a choice he made having been convinced of Gandhi's common sense approach to ameliorating the housing situation for the poor. His main contribution is in shining light on the path of resistance to the uncritical acceptance of colonial mindsets in post-colonial Indian architecture by persistent employment of methods emerging from both his Methodist Christian background, as well as his encounter with Gandhi. Sennett, exploring the phenomenon of the global foreigner, refers to two scar marks left on King Oedipus' body in Sophocles' play *Oedipus Rex*: 'the first wound marks his origins, the second his personal reckoning of his life. Twice wounded, he has become a man whose life can literally be read in his face' (Sennett 2002: 192). Yet, contrary to what Freud would have us believe, the second wound – the one that would not heal – opens an entire new chapter in the life of Sophocles' King Oedipus. The raw condition of the flesh resulting from the second scar suggests a continual cultural assault on the 'body' of the foreigner – both from within (self-denial) and without (self-injury) – an assault simultaneously painful yet rewarding in its revelation. The two scars – metaphoric lenses offering glimpses into the foreigner's lived-in experience – complement the interrupted experience of the foreign landscape with interludes of memories of belonging. These scars or windows – Baker's Christian beliefs and his acceptance of Gandhian principles – had turned him into an instrument of interface.

An introversion of the universal knowledge took place in Baker, alongside his emerging introspection of everything Indian, much like Franz Kafka's travelling salesman in his short story, *Metamorphosis*. Confined in bed in his own house finding himself one morning turned into a giant insect, he grows fearfully aware of – yet resistant to – his diminishing visual grasp on the blasé reality outside his window but increasingly familiar with the 'other' universe of reality presented through the picture on the wall. Baker's foreignness was reduced through a 'de-garbing' – both literally as he increasingly took to Indian ways of life and manner of dressing, as well as by operating as a 'primitive' practitioner whose architecture remained unpretentious. An interface had been opened up in which his British-ness was quite literally in dialogue with his embraced Indian-ness.

Thus in Baker's architecture the carefully conceived and detailed enclosure became the site of persistent dialogue between nature and human habitation, in which weathering and maintenance helped to sustain a dynamic equilibrium. A negotiation opened up between the deep-rooted practice of crafting in the vernacular building culture of Kerala and the desires and products associated with technological development. Perhaps Baker's buildings never left that 'primordial moment of figuration and becoming' that Deane talks about (Deane 2009: 146); foreground and background are eternally interwoven and what we perceive as foreground or background are, after all, momentary snapshots of this rhythmic sequence. Frascari, in attempting to understand the role of detail as generator of design, defines the double-faced presence of technology – as *logos of techne* and *techne of logos* – and emphasizes that '*construction* and *construing* of architecture are both in the detail' (Frascari 1996: 500). Baker employed detailing not only to achieve connection between different architectural components but as sites of dialogue – thresholds as interstitial sites for constructing relationships between tradition and modernity, technology and culture, past and present, the local and the global, and even between his acquired Indian-ness and his innate foreign-ness (Figure 4.6). Understanding of detailing as threshold – between space and technology, between *techne* and praxis – helps bring the recessive nature of technology out of concealedness. The role of technology in Baker's work, like his own 'foreignness' – latent yet selectively expressive – acted as a hinge between India's past and present and in foregrounding inhabitation.

Figure 4.6 Detailing in Baker's home, the Hamlet.

Chapter 5

Spatial memory and identity in the work of Istanbul/Vienna artist Canan Dagdelen[1]

Liane Lefaivre

Introduction

In an increasingly globalized world, identity is an issue that is becoming more central to our lives. For a Muslim in Europe this is particularly important.[2] It is a theme running throughout all the art of Canan Dagdelen (1960), born in Istanbul, now living in Vienna where she teaches ceramics at the University of Applied Art. *Spatial Memory/Uzamsal Bellek* is the title of her most recent exhibition at the American Hospital in Istanbul 2012.

Of all the forms of memory, spatial memory is the one that has most fascinated the molecular biologist Eric Kandel, who won the Nobel Prize for his work on the subject. One primary reason for this fascination, as he explains in his *In Search of Memory* (2007), which mixes autobiography with a history of his scientific field, is a terrible loss. As a young child, he was uprooted in the wake of his family's escape in 1938 to the United States from the anti-Semitism of Vienna that would eventually lead to the Holocaust. He has been haunted by this loss – of a time, of a world, of a people, coming together as a sense of place – all his life.

There are other reasons why spatial memory is especially important to Kandel. He points out that the idea that the ability to represent space is built into our minds is an old one; it goes back to Kant, who pictured people 'as being born with principles for ordering space and time, so that when other sensations were elicited – be they objects, melodies, or tactile experiences – they are automatically interwoven in specific ways with space and time' (Kandel 2007: 307).

Perhaps this deep cognitive importance is why spatial memories evoke such empathy. Olfactory, acoustic impressions, tastes, and textures elicit memories. But the opposite is not true. We cannot reconstruct them in our minds the way we can architectural settings. In addition, architectural spaces come back to us vividly in dreams in a way that sound and smell and taste and tactile impressions do not.

The idea that there is something innate in the human mind, something 'pre-wired' in our thinking about spatial cognition, and that there is even something architectural about how we remember, has very deep roots. It goes back to antiquity. As Kandel again writes, in 55 BC Cicero 'described the Greek technique of remembering words by picturing the rooms of a house in sequence, associating words with each room, and then mentally walking through the rooms in the right order' (Kandel 2007: 307).

There is something topological about memory. In other words, the reason architecture is experienced so vividly and is such an effective mnemonic device may be that our minds are capable of structuring memories the way architects order different spaces in buildings. A recent case points to just how effective architecture can be as a mnemonic device. Historian Tony Judt, paralyzed with ALS in 2010, used his memory of a much beloved chalet in the Swiss Alps in this manner in order to collect his thoughts in the last year of his life and managed to write two bestselling books (Judt 2011). Orhan Pamuk also famously used architecture as a means of holding onto memory, turning his *Museum of Innocence* (2009), about loss, into a real building full of the objects that his fictional self had collected. In the same vein, the philosopher Ludwig Wittgenstein, in his *Philosophical Investigations* (1963), used the metaphor of building a house – a task he knew firsthand as the architect of his sister's house in Vienna's third district – to describe how we build language and think. Aptly, it is in the foyer of the Wittgenstein house that Dagdelen recently presented her airy, lace-like installation *lace*, another reflection on memory and space, where the p in the English word 'place' is omitted.

The title of Dagdelen's exhibition is *Spatial Memory/Uzamsal Bellek* and is a direct reference to Kandel's book on memory, especially on spatial memory, which fascinated her when she read it (Figures 5.1 and 5.2). Dagdelen tends to explore the subjects of loss, space, architecture, and memory, as an artist from Istanbul resident in Vienna. Like Kandel's work, hers too, in its own way, is concerned with the process of learning, of how the brain captures and sustains the memory, and how all inputs are connected in the end. There are two words in Turkish to refer to space; she chose the word *'uzam'* for the title over *'mekan'*, because it is closer to Kandel's dynamic definition of memory that is something that is constantly evolving as opposed to something static and fixed.

One reason Dagdelen felt an affinity with Kandel's work on spatial memory is the primarily spatial nature of what she calls her *dot* sculptures, or what the nineteenth-century art historian Adolf von Hildebrand called 'effective form' (Figures 5.3 and 5.4). He distinguished two different kinds of form in sculpture: inherent form and effective form. Inherent form is conceived of as independent of its surroundings. In contrast, effective form is conceived in order to affect the onlooker's perception of the space that surrounds it (von Hildebrand 1893). These *dot* installations are mysterious, disconnected constructions illusionistically floating in space, on one hand as architectural messages, as it were, on the other as written words. From her earlier installations *AT HOME dot* (2004), to her Mahya design *Selâm* (2010), they all can be seen as examples of von Hildebrand's effective form. They reverberate in space and transform its perception, like the pointillistic,

ure 5.1 *Spatial mory/Uzamsal Bellek,* 1.

re 5.2 *Spatial ory/Uzamsal Bellek,* detail.
less mirror steel,
blasted, glossy
elain, silkscreen

35 x 0.2 cm.

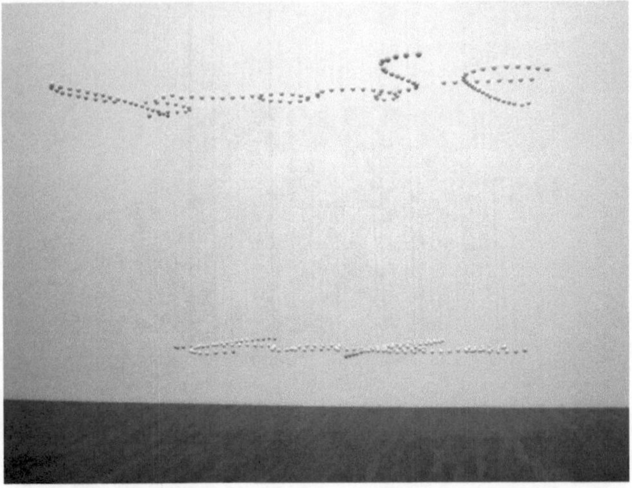

Figure 5.3 *POET dot GARDEN dot*, 2003–05
Porcelain, fine steel c
2 parts, together 300 x
160 x 400 cm.

Figure 5.4 *TABANI d*
2007.
Aluminium, copper, fi
steel cord
300 x 180 x 180 cm.

evanescent crowns of light that are invisibly suspended from the domes of Istanbul's mosques or the traditional Mahya wishes or prayers, lit up against the night sky between their minarets during Ramadan.

But the main reason for Dagdelen's empathetic response to Kandel is that her work is also about the spatial or architectural memories of Turkey, the world where she spent her childhood. These memories that she tries to recapture in her works are autobiographical, but they are also collective and typical of what she sees as Turkish culture as a whole.

This is why she is primarily interested in what she feels is the elemental form of Islamic architecture, the domed structure (Figure 5.5). In Dagdelen's work, this elemental shape – whether in *dot* form or as miniature solid structures, both lost in space – embodies the search for a place in a world that had become placeless.

ıre 5.5 *dome.*, 2005.
celain, nylon cord.
meter 62 cm, h 32 cm.

The famous architect of modern Turkey, Sedad Hakkı Eldem, a nephew of diplomat, archaeological pioneer, museum director, and orientalist artist Osman Hamdi Bey, was also concerned with the relation between East and West (Bozdoğan *et al.* 1987; Akcan 2008). His way of expressing this concern is typical of a certain strand of regionalism. He creates a synthesis between elements of tradition and elements of new international modern styles. There is something unusual about Dagdelen's choice to reinterpret Eldem's Ağaoğlu House of 1936 by her installation *OTHER dot/ÖTEKİ dot* (Figure 5.6). She chooses the architectural element that is responsible for shaping the different spaces in the building, the ground plan. What is remarkable about this particular ground plan of Eldem's design she has chosen is that it incorporates a large circular room which is typical of the traditional Turkish house, one in which the divan is placed and which is also the hall called 'sofa'. But whereas for Eldem the relationship was seamless between this traditional Turkish space and the rest of the modern functional plan, for Dagdelen it is not. In her version of the ground plan, the hall is disconnected and floats above the rest of the plan. The relation between the

Figure 5.6 *OTHER dot ÖTEKI dot*, 2010.
Metal, fine steel cord, mirror
190 x 175 x 5 x 123 x 5 c
Mirror diameter 50 cn

two spaces, the space of cultural memory and the space of the rest of the house, is seen as precarious. Their bond is unstable and on the verge of breaking. Implying a questioning of traditional expressions of Turkish identity in a subtle, indirect way, the two spaces are beginning to dissociate from one another, to break apart. In other words, she is using the technique of what the early-twentieth-century literary critic and novelist Victor Schklovsky called 'strangemaking'.[3] She is taking a juxtaposition of elements that has become an overfamiliar commonplace and she is *making it strange*. The effect is to make the onlooker rethink and perhaps revitalize this relation.

As Kandel says, without memory life would break into as many fragments as there are seconds in time. 'Memory is the glue that holds the mental life together and gives it continuity. As a result, we are who we are in large part because of what we have learned and what we remember from past experience' (Kandel 2007). It takes an effort of remembering to keep these structures together. When looking at her series of ceramic yurts ('yurt' means homeland and nomadic tent in Turkish) – *bir GÖZde*, *Homely*, or *Homewards* – it is obvious that they are saying the same thing. Because they have been uprooted from the soil of tradition, they have fallen apart. To put them back together has become difficult. The glue that binds them in such a forced, conspicuous way gives them an air of artificiality and precariousness.

The theme of home has undergone a change very recently in Dagdelen's work. In her most recent exhibition, entitled *Covalent Bonding*, the image of a dual home appears, bonded at the base. Gone is the fragility of the walls. The bonding of the two units is complete. The seams have disappeared. The fusion between the two homes is resolved. Is it possible to be at home in two worlds, in East and West, in two identities? The object answers in the affirmative.

But now Dagdelen has turned from ceramic to the much more solid medium of steel. In keeping with this new material, the bonding is firm and unbreakable. The disconnectedness of dots has been reversed. The two worlds have fused. The two similar archaic architectural forms become one through the use of steel, one of the prime materials of contemporary modern architecture. The image of uprootedness is still there. The home is still suspended in space. But like the earth turning on its axis, the two elements form a whole. This sculpture, entitled *Synapse* – coined from the Greek *syn-* (together) and *haptein* (to clasp) – is an image of cohesion (Figure 5.7). Again with this title, *Synapse*, Dagdelen has been inspired by Kandel. He explains that memory occurs as a result of synaptic learning in the brain which takes place by strengthening connections between neurons. So it is in Dagdelen's piece.

Remarkably, through its use of reflective polished steel, Dagdelen has made out this cohesive form, uniting two worlds, East and West, in which everyone can literally see their own reflection. This may be an image at variance with these times so fraught with fragmentation, but it is the kind of spatial memory that anyone can identify and connect with. And like any 'effective' sculptural form, one can only hope that it is capable of transforming the space that surrounds it and the minds that inhabit it.

re 5.7 *Synapse/ ps*, 2011. nless steel, matt, sy x 97 x 97 cm, total ht 250 cm. l base 298 cm eter, edition 1/3 +1.

Part Two

Memory and identity

Chapter 6

Psychoanalysis and identity in architecture

John Hendrix

Introduction

Architecture at its best is an expression and reflection of the human psyche, in which cultural identity plays an important role. In order to understand better the role that architecture plays in changing concepts of cultural identity, it is helpful to understand better the role that cultural identity plays in the human psyche. In the increasingly globalized twenty-first century, it is important for vernacular identities to be represented in architecture. At the same time, globalized technology unites everyone in the world in a singular identity. All cultures are also united by universal structures of human language and behavior, and it is important that this be represented in architecture also. This paper presents a model of analysis for architecture based in psychoanalysis, which proposes to equally represent individual and universal identities.

Language, imagination and cultural identity

Cultural identity is given by language, and an important part of the psyche is composed of the laws, relations and customs of a cultural identity as given by language, what Jacques Lacan called the Symbolic Ego, as given by the Other, the linguistic matrix, which is what forms the unconscious (Lacan 1977 [1966]: 1–7, Lacan 1968 [1956]: 47–49). The Imaginary Ego, the independent bodily and sense experience of the individual, is absorbed into the Symbolic Ego, the use of language, as a result of the Mirror Stage, in the childhood of an individual, though not without internal conflict (Lacan 1968 [1956]: 47–49). When a child sees its image in a mirror for the first time, its physical self-identity is reified as a body image, which becomes a signifier in the language of self-identity (Lacan 1977 [1966]: 1–7). After the Mirror Stage, it is impossible to have an immediate sense experience without the intervention of language. Sense experience is filtered through language and signification, including the experience of architecture.

In architecture, how can the physical, sensual experience of the architecture be absorbed into the symbolic, linguistic experience of the architecture so that the architecture reflects the role of cultural identity in the human psyche in a way that is resonant to everyone in a culturally pluralistic world? What is the relationship between architecture and individual and collective ego formation? In architecture, as in any language, particular vocabulary elements have particular cultural associations. This makes the architecture, in one way, metaphysical: the vocabulary elements function as signifiers, and communicate an idea which is independent of its material presence. How do the signifiers represent culturally specific ideas in the material presence of architecture in a way that has resonance for everyone? What are the elements of identity formation, as communicated in architecture, that all human beings share?

According to Lacan, the irrationality of a rational sequence of metonyms, the presence of absence within signification, or within meaning in rhetorical language, is made present at the point the metonymic chain in language produces signification, which is the 'anchoring point', the *point de capiton*, which occurs retroactively, after a phrase is completed, and is the point at which the network of signifiers in the metonymic chain corresponds to a network of signifiers in the concept and thus accomplishes signification (Lacan 1977 [1966]: 303). In other words, if I speak a sentence to someone, communication occurs when I say something that the person already knows or understands. Thus the communication is retroactive, and there is a gap between the word or signifier that I utter and the understanding of the word when it is heard. Ultimately, for Lacan, this model can never be ideally realized, and all communication involves misunderstanding. The metonymic chain accomplishes this model without 'crossing the bar' into meaning, or the signified; the idea is not present in the metonym itself. Communication occurs through the play of the signifiers, without a metaphysical element in language (in the relation between signifier and signified), except in the gap between signifiers in communication.

This type of production is labeled by Lacan '*signifiance*' (Lacan 1977 [1966]: 152–154, 316–317), as opposed to significance or signifying, a type of production in language which does not cross the bar, the point of resistance, between the signifier and the signified in language, from the Saussurean model (Saussure 1986 [1916]: 110–115). The relation between the signifier and the signified cannot produce a metaphysical element in language; it is the gap between them which is the metaphysical element. *Signifiance* contains the presence of the irrational, as in architecture, or any communicative system, and the absence of the signified in relation to the signifier, and the absence of the subject as defined by the signified. As the concept of the metonym in language corresponds to the process of displacement or *Verschiebung* in Freudian dream work, in the relation between dream thoughts and dream images, the dialectic of the rational and irrational in language corresponds to the dialectic of conscious and unconscious thought. A metonym involves a displacement in that it is a comparison with something that has been displaced, as in the foot of a hill, or the mouth of a river. This correspondence leads to the formation of the Lacanian concept of the unconscious, that the unconscious is structured like a language (Lacan 1981 [1973]: 20, 126, 149, 198), and it is through this concept that architecture and psychoanalysis, or the philosophy of the structure of the psyche, or the philosophy of the identity of the human subject, and in particular its identity in language, and the study of the mechanisms of conscious and unconscious thought, correspond.

Unconscious thought is seen as a different form of thought than conscious thought, and exhibits aspects of the irrational in relation to the rational. The linguistic structure of the dream image in Freudian psychoanalysis is seen as diffuse, clumsy and awkward; it is missing the organization of conscious reason, while its forms are mimetic of it. If the unconscious is the discourse of the Other, in Lacanian terms, if the unconscious is the network of language and rules into which the subject is inserted, as Lacan claims, it is only so insofar as it is a mimesis of the discourse of the Other (Lacan 1981 [1973]: 130–131, 144, 198). Dream images, the manifest content of the dream, as defined by Freud in *The Interpretation of Dreams* (Freud 1965 [1900]: 340–343, 346–349, 544–545), are not capable of representing logical relations between the dream thoughts, the latent content of the dream, or of representing logical relations between conscious thoughts, the relations created by syntactical rules in rational logic.

Dream images can be compared to architecture and the visual arts in their relation to the syntactical structures of language. The desire on the part of the visual arts, in particular architecture, to engage as much as possible the syntactical structures of language reflects the desire on the part of the arts to interweave the Imaginary and the Symbolic, in Lacanian terms, the world given by the senses and the world given by language, in the complete constitution of the subject. The Imaginary corresponds to the perception of the image, and the mnemic or memory residue of the image which is incorporated in dreams, while the Symbolic corresponds to language, the signifying structure or Other into which the perceiving subject is incorporated. In Freudian dream analysis, dreams are a function of both the Imaginary and the Symbolic, as compositions of images which do not correspond directly to a logical syntactical structure, but the composition of which is in fact determined by underlying linguistic structures in unconscious thoughts, what Freud called the latent content, which constitute the structure of the unconscious (Freud 1965 [1900]: 311–313).

Linguistic structures found in dream composition include metaphor and metonymy. Lacan demonstrates that these reveal the unconscious in conscious discourse (the irrational within the rational). Linguistic structures in dream composition also include syntactical mechanisms such as ellipsis, pleonasm, hyperbaton, syllepsis, apposition, catachresis and antonomasis. These are the mechanisms of condensation and displacement, to use Freud's terms, regression and repetition, that contribute to the composition of dream images as elements of the unconscious registered in the subject by the Other. The Other is the linguistic superstructure that determines the subject in ways that are not always conscious, and that do not always correspond to rational thought, but can be revealed through dream work and compositions in art and architecture. In that these are the mechanisms which allow the subject to moderate the dream discourse, they are not mechanisms of the dream itself, but rather the conscious reading of the dream; the unconscious can only be found in conscious thought as an absence.

Lacan would thus conclude that the unconscious does not exist other than as an absence in conscious thought (Lacan 1981 [1973]: 144).

Among the syntactical mechanisms, ellipsis involves the omission of a word in a syntax without altering the signification, introducing a gap in the syntactical structure. A pleonasm is the use of more words than are necessary for the signification, so it is a repetition, and a condensation, as in metaphor. A syllepsis is the use of a word in a syntax which agrees with one word in the syntax grammatically but not another, so it is a displacement, as in metonymy. An apposition is the placing of a word in a sentence to explain another word, as a repetition. A catachresis is the incorrect use of a word; an autonomasis is the use of a title instead of a name. These are the primary condensations and displacements, along with metaphor and metonymy, which are active in both language and dream construction, according to Lacan. According to Freud (Freud 1965 [1900]: 311–313, 330), condensation and displacement are the primary mechanisms in the transformation of dream thoughts into the dream images. Words are either unconsciously combined in ways that are transferred to unrecognizable images, or they are displaced. All of the syntactical structures can be used as strategies in architectural composition.

Syllepsis can be found in the overlay or interweaving of geometrical forms, for example, where a form which is rotated or shifted might correspond to one underlying form in a certain way but not to another. Such an overlaying can result in the creation of a 'transformational relation' (a term from Chomskian linguistics, applied to architecture by Peter Eisenman), which signifies the relation between the form of the composition and the underlying conceptual structure of the composition, the surface structure and the deep structure (also from Chomskian linguistics), or the signifier and that which is signified (from Saussurian or Structural linguistics). The transformational relation is comparable to the *point de capiton* of Lacan. As Vitruvius wrote in *De architectura* (I.1) in the first century BC, 'Both in general and especially in architecture are these two things found: that which signifies and that which is signified' (*quod significatur et quod significat*) (Vitruvius 1931 [27 BC]: 7). That which signifies is the *verba*, or words in rhetoric, the material vocabulary of architecture, and that which is signified is the *res* (proposed thing, relation).

Language, imagination and architecture

The signifier of the architecture is the form (*forma, eidos, idea, species, imago*, all terms used by Vitruvius and his contemporaries), geometrical solid and architectural vocabulary type, and their sensual appearance in space as perceived by the viewer, as in the Imaginary of Lacan. The signified is the conceptual relations between the forms as they are constructed in the mind of the viewer, as in the Symbolic of Lacan. The signified in architecture, the conceptual relations in the mind of the viewer, involves concepts such as frontality as opposed to spatial recession, solid as opposed to void, symmetry as opposed to asymmetry, axial as opposed to centrifugal; in other words the disposition of the forms as conceptualized in the composition, in the same way that the signified in language involves condensation and displacement, the syntactical mechanisms of dream work. In linguistics, the signifier of a language is the phonetic form or physical signal, while the signified is the corresponding mental analysis that arises in response to the physical signal of the phonetic form, as in the reading and experience of architecture.

An example of the overlaying, rotating and shifting of geometrical forms, producing transformational relations, enacting a dialectic between the physical presence of the building, the signifier, and the conceptual organization of the building in the mind of the viewer (involving frontality and recession, solid and void, symmetry and asymmetry, axial and centrifugal) can be seen in Giuseppe Terragni's Casa Giuliani Frigerio (1939–1942, Figure 6.1) in Como, Italy. Peter Eisenman presented such an analysis of the building in 'From Object to Relationship II: Giuseppe Terragni, Casa Giuliani Frigerio', in *Perspecta* 13, The Yale Architectural Journal, 1971.

The syllepsis, where a word agrees grammatically with one word but not another in a syntax, can act as the *point de capiton*, Lacan's anchoring point in the signifying chain, which is the point at which the overlay and interweaving of the words as signifiers or forms reveals an underlying conceptual structure, through syntactical mechanisms, as in the transformational relation. The anchoring point of Lacan is the point at which, in the retroactive anticipation of meaning on the part of the subject in the course of the diachronic establishment of a syntactical structure, a sentence for example, the subject enters into the structure in the gap between signifiers, linguistic units, and is represented by one signifier to the next signifier, either diachronically or retroactively, and the subject becomes aware that there is a distinction between the signifier in language and itself, creat-

re 6.1 Giuseppe agni, Casa Giuliani rio (1939–1942), o, Italy.

ing a bar which cannot be crossed, because the structure of the signification, the discourse of the Other, is the unconscious, which is not accessible by conscious thought.

The anchoring point reveals the presence of the unconscious in conscious thought, and it reveals the presence of the irrational in rational discourse. The anchoring point can be analyzed in the syntactical mechanisms of dream work, or in architectural composition, for example the point of the transformational relation between signifier (form) and signified (concept). This can happen when two or more syntactical organizations (grids, for example) are combined and then shifted or rotated, or forms are condensed or displaced, revealing the signified of the composition in the signifier, or, as it might be seen, the unconscious in conscious thought.

An apposition, or repetition in signification, might be one element of a geometrical form which is designed to correspond to another geometrical form in an architectural composition, revealing the mechanisms of the signification. The apposition may be the doubling of an element, a self-repetition that reveals the repression of another element, as in metaphor. A catachresis, the incorrect use of a word, might be that element in an architectural composition which is purposefully out of place, as often happens in the dream; the catachresis reveals the self-enclosure of the signifying system of the language, and the production of the irrational in rhetorical language, as in metonymy. An autonomasis, replacing a word with a name or title, might be the substitution of a type-form for a form in the architectural composition, to stage the dialectic between the metaphysic of the architecture and the *signifiance* in which it is engaged, the signifying production in its syntactical organization, in the dialectic between the Imaginary and the Symbolic, the signifier and the signified.

The coexistence of the mnemic or memory image and the phonetic element in the writing of a dream in Freudian psychoanalysis is the coexistence of the Imaginary and Symbolic in Lacanian psychoanalysis, and the coexistence of the mnemic residue of the visual perception and the mnemic residue of the auditory perception, the traces interwoven into the language of the unconscious. The visual residue is the 'thing presentation' or *Sachvorstellung*, according to Freud in *The Interpretation of Dreams* (Freud 1965 [1900]: 374–377), and the auditory residue is the 'word presentation' or *Wortvorstellung* in the formation of the dream image, which is described by Freud as the transition from the latent content to the visual image of the dream in a 'concern for representability', or *Rücksicht auf Darstellbarkeit*. The coexistence of the thing and the word in the representation, in the writing of the dream, is a 'double inscription' or *Niederschrift*, which involves condensation and displacement, repression and repetition, and which corresponds to the coexistence of conscious and repressed or unconscious images which may occur in the preconscious, in the memory of the dream, and which constitutes the structure of conscious language in the mechanisms of metaphor and metonymy in particular (Lacan 1977 [1966]: 158–164).

The *Niederschrift* contributes to the *Vorstellungsrepräsentanz* of the dream, which is the representation of a representation, the dream image representing the perceived image (which is already a representation in Kantian or Platonic terms), according to Freud. According to Lacan (Lacan 1981 [1973]: 60, 217–218), the *Vorstellungsrepräsentanz* is actually that which takes the place of or substitutes for the representation. The mnemic residue of an image in perception is transformed by syntactical mechanisms in the unconscious as it is seen as the mnemic residue of the dream. The replacement of the representation poses a distinction between what is seen by the subject and what is represented in the mind (as in the distinction in classical philosophy between *morphe* and *eidos*, or in Scholastic philosophy between *species sensibilis*, sensible form, and *species apprehensibilis*, intelligible form). The distinction is between the signifier and the signified, as given by language, which structure perception itself (in the sense of Kantian a priori intuition). The distinction between what is seen and what is represented to mind is the disjunction between sense-experience and reason, or the sensible and intelligible, in metaphysics, between the real and the ideal in Transcendental Idealism, or between the Imaginary and Symbolic in Lacanian terms.

The disjunction of the metaphysic is displaced, in linguistics and psychoanalysis, from the structure of reason in consciousness to the structure of language and the unconscious, as seen in the *Vorstellungsrepräsentanz* of Freud, the representation of the representation. The distinction between what is seen and what is represented reveals a divided subject, in Lacanian psychoanalysis, a subject which does not know itself, or the mechanics of its thought, and a subject which is divided from its conscious reason, as reason becomes other to itself, as the unconscious mechanisms of reason are inaccessible to conscious thought. The basis of Lacanian psychoanalysis is to be found in linguistics, and in the study of perception, the nature of representation in vision. It is for that reason that psychoanalysis has such importance for art and architecture, and can be helpful in addressing questions about cultural identity in architecture. The answers can be found in particular in the elements that all architecture shares, and the ways in which all human beings interact with architecture.

As a result of the complex network of psychical relationships which produce the dream images, and the mechanisms of condensation and displacement, dreams are composed of disconnected fragments of visual images, syntactical structure in language, and thoughts, the *Sachvorstellung* and the *Wortvorstellung*, which are seen in a variety of logical relations to each other in a palimpsest of traces which is difficult to unravel, and which can appear to be irrational, according to Freud (Freud 1965 [1900]: 318, 340–343, 346–349, 353, 544–545). Architecture is often experienced in the same way, as a composite of individual perceptions or experiences; architecture can appear as an irrational combination of a palimpsest of layers of rational relations between traces. The palimpsest of dream images can be seen as the condensation and displacement of figures and spatial relationships, such as foreground and background, and the coincidence of opposites, as in the simultaneity of frontal and recessional or solid and void, in a 'phenomenal transparency', to use Colin Rowe and Robert Slutzky's term (Rowe and Slutzky 1963: passim), or conceptual organization of the architecture on the part of the viewer.

An example of the palimpsest of layers of rational relations between traces, involving the condensation and displacement of forms, can be seen in Peter Eisenman's Wexner Center for the Arts, Ohio State University (1983–1989, Figure 6.2). The Wexner Center also combines the Imaginary, in the form of the fragments of building types previously associated with the site, and the Symbolic, in the form of the gridded matrix placed on the site. The Imaginary, as fragments of traces in a palimpsest (as in conscious reason), is absorbed into the Symbolic, into language and the Other, as the conceptual organization of the building. Freud saw the mind as a palimpsest, comparing it to the city of Rome, where the city is composed of fragments of traces that have floated to the surface from history, or from the subconscious mind (Freud 1994 [1930]: 6).

The network of logical relations that contribute to the composition of dream images is too complex to be unraveled in dream analysis, according to Freud (Freud 1965 [1900]: 318, 340–343, 346–349, 353, 544–545). Displacement, condensation, fragmentation, substitution and the *coincidentia oppositorum*, coincidence of opposites, are products of the complex network of logical relations, or the mnemic residues of such, in the *Vorstellungsrepräsentanz* in dream thoughts, which is too complex to correspond to any logical structure. In the process of the dream formation, the logical links that hold the psychical material together are lost. It is the task of psychoanalysis to restore the logical connections which the dream work has destroyed, as dreams are seen as access to a knowledge of the unconscious activities of the mind, as Freud wrote in *The Interpretation of Dreams* (Freud 1965 [1900]: 346–349, 647), an access to the psychical mechanisms which psychoanalysis seeks to understand. Lacanian psychoanalysis furthers this quest in the analysis of the linguistic mechanisms of which dreams are a product.

Figure 6.2 Peter Eisenman, Wexn Center for the Ar (1983–1989), Ohic University.

A psychoanalysis of architecture can reveal the presence of the Other, the linguistic and cultural matrix which constitutes the identity of the subject and its unconscious thought, the Symbolic, in relation to the self-identity of the subject as an individual body, or the Imaginary, and can reveal the mechanisms by which architecture can contribute to cultural identity in a way that both reinforces cultural identity and at the same time allows the architecture to be resonant to all human beings, without being exclusively culturally specific. In this sense a psychoanalysis of architecture is a continuation of the utopian modernist goal of producing an architecture which can communicate to anyone regardless of cultural background, from the transcendental forms of De Stijl to the appropriated forms of the New York Five. A psychoanalysis of architecture would establish more developed relationships between architectural forms and the individual or cultural psyches which they reflect and represent.

An example of a building which combines a culturally specific identity with a universally understood structural relationship between the Symbolic and the Imaginary, the matrix of the Other and the physical presence of the building, is Le Corbusier's Villa Shodhan (1951–1956, Figure 6.3) in Ahmedabad, India. The building displays the same dialectic as Terragni's building, between the conceptual organization and the material presence, as enacted by the overlaying, shifting and rotating, to produce the same relationships in the signification of the building, between frontality and recession, solid and void, symmetry and asymmetry, axial and centrifugal. The formal composition is combined with elements that formed the theoretical basis of the architecture of Le Corbusier's late career: the *béton brute* (rough-formed concrete), *brise soleil* (angled wall to take advantage of the sun), and landscaped terrace (to connect the villa to its surroundings). These elements were developed by Le Corbusier after

re 6.3 Le
usier, Villa
dhan (1951–1956),
edabad, India.

World War II to produce a more humanistic, and less positivistic, less machine-oriented architecture. The psychoanalytic reading of the building promotes the agenda of the Surrealist movement, which also influenced Le Corbusier and contributed to his humanistic architecture. In combination with all of this, the villa displays the indigenous patterns of forms and living spaces associated with its surrounding region, thus absorbing a local cultural identity (as the Imaginary) into a structuralist psychoanalytic matrix (the Symbolic), combining local and global identity structures.

The relation between the dream image and the dream thought in dream work, and the signifier and signified in any language, can be seen in the relation of the subject to language. The dream image responds to the dream thought, the latent content, in the *Vorstellungsrepräsentanz*, as a form of psychical activity in response to perceptual activity. The content of the perception is anticipated and rearranged, as the subject is anticipated in language, retroactively; the word represents the image to another word as the signifier represents the subject to another signifier, and it is that series of relations which make both the dream and language intelligible. The representation of the image by the word, or the subject by the signifier, are tools for intelligibility and communication in architectural composition.

Lacan compares the distortion or *Enstellung* described by Freud in the dream work to *glissement* in signification (Lacan 1977 [1966]: 160), the sliding of the signified under the signifier in the course of the signifying chain, as seen especially in metaphor, where one word takes the place of another, which bars the subject from signification, from the language which it produces in conscious thought. In the gap between what is perceived and what is represented in the dream as the mnemic residue of perception, a direct connection is lost in the process of distortion, as the connection is lost between the signifier and the signified in language. Freud's dream condensation or *Verdichtung* is compared by Lacan to the combination of signifiers in metaphor. Displacement or *Verschiebung* is compared to the transfer of signification in metonymy, where the correspondence between signifier and signified is maintained, but shifted, and rendered nonsensical in rhetorical language, barring the subject from itself (Lacan 1977 [1966]: 160).

The only difference between the mechanisms in language and the mechanisms in the dream work, according to Freud, is the difference between the intentions of communication in language and the consideration of representability in the dream, which is also a mechanism of a type of communication, combining both *Wortvorstellung* and *Sachvorstellung* in conscious discourse, as can be seen in an architectural composition in the combination of the syntax and the form. The dream has no intention of communicating anything, according to Freud. The elision of the subject in language, in the dream and conscious discourse, creates an absence of the subject to itself in its reason. The anticipation of the subject in the signifying chain caused by the absence of the subject, which occurs at the *point de capiton*, or the inaccessibility of the unconscious, is that which causes desire in the subject in signification; the impossible attempt to find fulfillment. The desire of the subject is the desire of the Other, for Lacan, the discourse of which the unconscious is composed, and which is always present in architecture or any human activity. Desire is enacted by the *objet a* of Lacan, that which is missing from the subject. In perception the *objet a* is found in the Gaze, that which is missing from perception.

Architecture – in particular in compositional strategies and the interweaving of geometries, devices which can be found in any culture – enacts the *objet a* and the Gaze in describing the desire of the subject, in relation to the Other (the cultural matrix), in the division of the subject in psychoanalysis and in the inaccessibility of the subject to that which defines it: the unconscious. In this way, architecture can function as a diagrammatic model of human identity. Architecture represents the most complete expression of human identity precisely because it entails the impossibility of the reconciliation of the subjective and objective, of form and function, of conceptual and empirical reality. Human identity entails the impossibility of the reconciliation of the conscious and unconscious, rational and irrational, Imaginary and Symbolic. In that architecture must always be tied to functional requirements, it is only art, that is, it expresses the human spirit, when its form contradicts itself, is other to itself, or when its form contradicts its function. A transformational relation is enacted, and the disjunction between conscious and unconscious thought, between presence and absence in the human condition, is revealed. This is true in any culture.

Examples of buildings wherein the form contradicts the function can be found in every period of architectural history, from the Parthenon in ancient Greece, the Pantheon in ancient Rome, the Gothic cathedral, the Renaissance palazzo and the compositions of De Stijl, Terragni, Le Corbusier and Eisenman. The design of the Pompidou Center (1972–1976, Figure 6.4) in Paris, by Renzo Piano and Richard Rogers, places the structural and functional elements of the

'e 6.4 Renzo
ɔ and Richard
ɪrs, Pompidou
er (1972–1976),
.

building on its exterior, subverting the traditional relationship between the Imaginary signifying aspect and the Symbolic signified aspect, between the physical function of the building and the formal or conceptual organization of the building (the form contradicts the function). The structural and functional elements displayed on the exterior are in excess of the requirements of the building; the building thus exhibits the excess production of late capitalism, and global capitalism, in Western culture. The building thus combines the psychoanalytic element of architectural composition with the contemporary cultural psyche or zeitgeist; it combines the individual unconscious with the collective unconscious of a global culture in its signification.

Conclusion

The unconscious of the individual, and the collective unconscious of a culture, are known as an absence within a presence, and architecture functions to reveal that absence, to reveal the unconscious of a culture, the zeitgeist of a culture, through a transformational relation, and thus communicate a cultural identity without being culturally specific. Such is an architecture in the age of evolving globalization. Globalization creates opportunities for universally shared ideas and identities that can be represented in architecture, in combination with unique cultural and individual identities. In this way architecture in the age of evolving globalization can contribute to the enrichment of human life and culture.

Chapter 7

Architecture and memory

Berlin, a phenomenological approach

Ana Souto

> Remembering can, and does, leave its mark on the skin of the city.
>
> (Jordan 2006: 1)

Introduction

This chapter explores the relationships between architecture and memory, based on the understanding that architecture is a clear manifestation of material culture, 'buildings and cities are instruments and museums of time. They enable us to see and understand the passing of history' (Holl, Pallasmaa and Pérez-Gómez 2006: 31). Architecture and the built environment are perceived as palimpsests, but without the context surrounding these structures; can we actually read them as documents of the past? Moreover, can a phenomenological approach to architecture enhance our understanding of the built environment?

This chapter uses a methodology based on the analysis of the literature on the subject, fieldwork and case studies with the intention to discuss the validity of the built environment as a 'museum of time'. The literature that has been produced on Memory is overwhelming; as a consequence, for the purpose of this chapter, the most representative authors have been chosen. The last part of the chapter is devoted to the analysis of two case studies: the Topography of Terror and the Holocaust Memorial, both in Berlin. The analysis has been conducted using the theories explored in the first two sections, and the notes gathered from fieldwork in Berlin in July 2012.[1] The latter was designed to record information about different memory locations in the city, to identify how relevant the idea of authenticity or prosthetic memory was as part of the visitor experience.

On the one hand, this study was based on the understanding that authenticity should have the ability to legitimize a site as a repository of memory, and as such it should enjoy a certain aura that is not reproducible – borrowing Walter Benjamin's theory on art (1968). Following this idea, it should be possible to feel this aura, the weight of authenticity. On the other hand, contemporary design of sites of memory has introduced special effects through light, materials and circulation with an intention to encourage 'experiential learning, emotional responses, [in order to] empathize and identify with people from the past' (Andermann and Arnold-de Simine 2012: 4). As a consequence, it seems that the experience gained in these locations does not necessarily depend on the authenticity of the place any more. This chapter aims to understand if authenticity is a significant element in the design of centres of memory, and if it encourages not only the understanding of a particular episode in history (in this case the Holocaust), but also a more complete experience of a particular site.

Memory, collective memories, identity

Memory and its collective and social dimension are intimately connected with the way we understand our built environment. Urban and natural landscapes act as reminders of the past and its hidden memories. Halbwachs stressed how collective memories are essential when reconstructing the past in the context of the present: even though different communities will transform and distort the past, they will always maintain some sort of anchor to former episodes, in an attempt to validate their positions (1992: 34).

Memory and collective memories have a very changing nature since their ultimate goal is to represent a variety of distinctive and even opposing identities (Kattago 2001: 2). By assuming the malleable nature of memory, identity has to go through similar processes of 'artefact, invention and social engineering which enters into the making of nations' (Hobsbawn 2003: 10). The assumed identity, after a process of selecting memory, highlights the most defining episodes and erases certain periods in order to construct a coherent narrative (Misztal 2003: 133). Selection, distortion and destruction are essential tools in the process of memory and identity construction (Passerini cited in Radstone 2000: 9); and the built environment, through its voids and reconstructions, becomes a palimpsest of these contradictory stories which support the manipulation of memory and identity.

A review of literature on memory shows the different ways in which societies have dealt and are currently dealing with the past:[2] in some cases the approach is a mere selection of certain episodes; in other instances periods of time are reduced or erased; in particular contexts, the decision is to reinforce or create new connections with the present. Despite the different approaches, there is always a clear element at the core: that memory '[. . .] is the essential condition of our cognition and reflexive judgement' (Misztal 2003: 1). The audience is ultimately the main agency behind memory, deciding what should be remembered or dismissed, irrespective of the efforts to materialize the past in memorials or museums.

The fall of the Berlin Wall in 1989 and the collapse of communism in Eastern Europe in the 1990s opened up a new and influential period in memory and its representation in the built environment. These events, together with the new political situations in South Africa and several Latin American countries (i.e. Argentina or Chile), forced the acknowledgement of traumatic episodes that had become paramount in the creation of new national identities. Difficult memories are now at the core of national representations with the hope of preventing these episodes being forgotten or dismissed as uncomfortable truths (Misztal 2003: 142). Keeping those memories alive could offer an opportunity to reconsider the past, confront it and, in theory, come to terms with it – a process that in many cases can be very controversial and painful (Edkins 2003: 2).

In this context, architecture and memory are strongly linked to the construction and destruction of identity; architecture, as material culture which represents and/or serves as a reminder of significant historical episodes, has been promoted to the category of heritage and as such should be protected and conserved. On the other hand, and as a consequence of the above, architecture has become the target of destruction which, on several occasions, has attempted to erase cultural identities (Bevan 2006: 16). This destruction of the built environment not only eases and promotes the introduction of new values and order; it also encourages the forgetting of the past, of collective memories which formed part of previous identities (Bevan 2006: 95, 174).

Memory and the built environment

Architecture, as material culture, plays a prominent role as a reminder of the past in the built environment, in the shape of memorials, museums, and traces; artefacts that embody the narratives of past events. Over the past few decades it has become apparent that events that damaged not only people, but the reputation or representation of a national identity, have been chosen to be remembered through architecture. WWII, the Vietnam War, 9/11 and the terror attacks in Madrid and London in 2007 have become part of not only shared histories and experiences, but also of landscapes. Nowadays, the aim of memorials, museums and traces is to include conflicting memories (Williams 2007: 174), and draw a tangible past into our present in order to understand it and learn from previous experiences (Bevan 2006: 4).

German identity was profoundly shaken before, during and after WWII, having clear consequences in the attempt to create a modern, post-war identity independent from the Third Reich (Young 1993: 22). The built environment did not help in this process, since the physical reminders of the Nazi past were more than evident. As a consequence, in both Eastern and Western Germany, the traces of the past were 're-appropriated, neglected, demolished or represented', following controversial debates on the approach to memory (Fullbrook 1999: 26). Sometimes, neglect and disuse were enough to erase certain traces from the landscape but in others time was only encouraging the conservation of certain Nazi monuments, such as the Nazi Rally Grounds in Nuremberg (Figure 7.1). In the 1970s, the new Bavarian state law on monument conservation stipulated that these monumental buildings should be conserved due to their 'historical

Figure 7.1 Documentation Cen at Nazi Party Rally Grounds Nurember refurbished by Professor Günther Domenig (1998–200

significance of its fascist "Kolossal" style of architecture, and their significance as "witnesses of the past"' (MacDonald 2006: 17). In order to avoid the ruins becoming mythologized, a Documentation Centre was introduced in the main building; it displays 'history' – aiming for a neutral and more academic position, whilst a museum showcases 'heritage' – artefacts and other elements worth retaining and protecting (MacDonald 2006: 22). The refurbishment designed by Günther Domenig – by leaving the main structure almost untouched as a ruin, punctured by the new structures introducing a new point of view in the exhibition and the Nazi past in general – follows a clear intention to introduce an emotional experience through the centre, rather than a mere educational one.

It has been argued that our ability to remember in a stable manner is strongly connected to the existence of cultural devices which support our understanding of memory. Objects record information in a way that surpasses the individual experience. However, without knowledge, we are unable to interpret the data that they store (Kwint 1995: 2). In order to make sense of our material culture, it is necessary to create an 'open dialogue between the object, the maker and the consumer' (Kwint 1995: 3), but so very often there is not enough information about the object itself, about the aim of the agency that created it, or who was to be the original consumer. Whenever any of these elements is either forgotten or transformed, the original meaning of material culture might be ultimately lost. The transmission of meaning can only be effective if the audience can actually understand the ultimate meaning of these objects, the context in which they were created, and the consumers they had in mind for their use (Cubitt 2007: 140).

Memorials

A monument/memorial 'in its oldest and most original sense is a human creation, erected for the specific purpose of keeping single human deeds or events (or a combination thereof) alive in the minds of future generations' (Riegl 1982: 21), a definition later broadly accepted and used. However, more importantly, what Riegl introduced in his original 1903 essay was the distinction between intentional and unintentional monuments, and the defining nature of the modern approach (cult?) to both of them: 'In the case of the intentional monument, its commemorative value has been determined by the makers, while we have defined the value of the unintentional ones' (Riegl 1982: 23).

WWII, and especially the fall of the Berlin Wall, introduced a new understanding of history and memory, drastically different to that of the nineteenth century. As opposed to memorials in which the narratives emphasized a glorious episode from history, triumphs or martyrdoms, contemporary memorials, monuments and museums through governments and private initiatives tried to come to terms with difficult pasts, hoping to transform collective memory through commemoration (Jordan 2006: 26). Collective memory has acknowledged different voices, and as a consequence new narrations are memorialized, like the role of the perpetrators (Young 1993: 21). Contemporary artists are dealing with the representation of troubled memories and events they themselves have not witnessed – narratives that have been influenced by layers of hyper-mediated experiences of memory (Young 2000: 1).

Moreover, and even when an artist finds the aesthetics that can support such a narrative (like minimalism instead of the narrative model), there is always the risk of a monument blending with the built environment, becoming invisible. Carrier points out how monuments can equally support memory and oblivion since they absorb the moral duty of remembrance (2005: 1), a concern that Young shares by arguing that 'once we assign monumental form to memory, we have, to some degree, divested ourselves of the obligation to remember' (2000: 94). Elsewhere, Young asserted that the best possible way to keep memory alive would be debating it constantly: through competitions, the public opinion and the media (Young 1993: 81). There is always a risk of trivializing history through these memorials (Edkins 2003: 126) – e.g., pieces of the Berlin Wall sold as a souvenir.

Spontaneous shrines, on the other hand, invite a more current and personal response to memorialization: small and perishable items carry very solid messages of solidarity, even though they tend to disappear alongside the currency of their episode, or are transformed into more permanent structures. These spontaneous shrines, 'the voice of the people' (Santino 2006: 13), are emotional, private and public at the same time, local and global, but most importantly, these shrines are non-institutional. And yet, they can become extremely popular and invite large numbers of people to behave in similar ways (e.g., Buckingham and Kensington palaces after Princess Diana's death). More often than not, these spontaneous shrines are removed in order to be institutionalized in monuments, plaques or artefacts that, ultimately, will be divested of their strong commemorative power.

Traces

Traces are 'indications of something that has disappeared, but has nevertheless left its marks and therefore in some sense still prevails' (Arnold-de Simine 2005: 12). These remains can be erased, preserved or revived, depending on the nature of the traces, the context in which they originated and were rediscovered, and, ultimately, the political will behind the rediscovery or erasure of the ruins. When reading the built environment as a palimpsest, we should try to identify how different societies and cultures have overwritten each other, selecting what should be visible, covered or erased from the surface (Eagleton 1981: 32). Traces can range from simple scars or ruins to the accumulation of new meanings and uses to pre-existent structures. Even though without clues related to the context it will be difficult to understand the events that these traces refer to, one could also substantiate the relevance of the concept of authenticity: these remnants are proof that something actually happened there.

Museums

Traditionally museums have been containers – receptacles where objects were exhibited, organizing memory around different artefacts (Misztal 2003: 21). Nowadays, museums try to showcase a selection of artefacts that has a meaning as a whole – certain coherence in the collection, or a latent symbolic or political message. Museums can offer the possibility to master the past (Maier 1997: 136), to give it a material form in a sequenced narrative in order to make sense as a unit. Walsh argues that one of the main reasons for the origin and development of museums, and their relevant role as keepers of heritage, was 'the desire to maintain the only things that nations can still call their own', since it emphasized the differences with other countries and cultures (Walsh 1992: 52). Heritage is then something irreplaceable: the inheritance of cultural continuity, that should be documented, respected, and safeguarded in order to facilitate its transmission to future generations (Matero 2000: unpaginated).

In the last few years a new phenomenological approach has been introduced into museology, which is based on the experience of the audience rather than on the objects exhibited, becoming 'more theatrical and more self-consciously rhetorical' (Hein 2000: 5). Since experience is at the heart of the visit, the object exhibited only serves as a prompt, so it does not necessarily have to be authentic, and as a consequence, has lost the aura that used to surround it (Hein 2000: 8). Memorial museums have an especial interest in reproducing history through their exhibitions, almost in an attempt to gain a cathartic experience. However, is it enough to evoke former times or particular events, or is it more important to be more accurate with history? How relevant is the concept of authenticity (Williams 2007)?

James Freed, the architect of the Holocaust Memorial Museum in Washington, visited concentration camps in Europe in order to gain inspiration for his design; he realized that this kind of museum could not be a simple exhibition of objects and artefacts – the museum had to 'reflect the moods of anxiety, fear and despair that enveloped the Jews of Europe during the Holocaust' (Ayer 1994: 15). By doing so, Freed started a new aesthetic in the museum: promoting a more emotional rather than mere educational experience. Examples of this new museology could be the two Berlin museums: Daniel Libeskind's Jewish Museum (1999) and the Workshop for the Blind in Museum Otto Weidt (in which phenomenology and authenticity are combined). These museums and memorials combine the search for truth (documentation and research centres) with the task of remembrance, always with an attempt to gain some closure with the event commemorated (Edkins 2003: 176).

Experiencing memory in Berlin

This section will be devoted to the analysis of two centres of memory in Berlin: Memorial to the Murdered Jews of Europe (The Holocaust Memorial) and the Topography of Terror (Figure 7.2). Based on the fieldwork undertaken in Berlin in July 2012, it will also apply the theories on memory explained in the previous sections. While the Holocaust Memorial is both a monument and a documentation centre that aims to commemorate the murdered Jews of Europe, the Topography of Terror is a trace which frames a new documentation centre, devoted to inform about the events that led up to the Third Reich and ultimately the Holocaust and WWII – both serve as reminders of the same episode, but their narratives differ in both form and approach.

The Gestapo Headquarters were demolished in 1949; not much attention was given to this site, which was further segregated from Berlin's city centre when the Wall was erected in 1961. It was only in the early 1980s that the site was 'rediscovered' and chosen for a memorial with associated areas of recreation. The winning design was ultimately dismissed, and the site was forgotten until 1985, when the 'Active Museum of Fascism and Resistance in Berlin' introduced memorial activism – the audience was encouraged to dig on the grounds of the former Gestapo Headquarters. Afterwards, a temporary exhibition was introduced on the site, emphasizing the authenticity of the place (Young 1993: 86–89).

In 1994 politicians of the Free Democratic Party (FDP) enquired about the possibility of locating the Holocaust memorial on this site, in order to highlight the authenticity of the place. However, this proposal was perceived as inappropriate and ultimately dismissed (Jordan 2006: 50). The site was finally transferred to the Topography of Terror Foundation, which organized a competition to build a documentation centre in 1993, choosing Peter Zumthor's scheme. However, the budget kept increasing and in 2004 it was decided to cancel the project and demolish what had already been built (Jordan 2006: 51) instead of leaving yet another layer in the history of the site. A new competition was held in 2006, which German architect Ursula Wilms won and finally built a documentation centre, which opened in 2010.

The analysis of these two centres of memory will be based on the ideas of Peter Carrier, who explains how a monument is linked to different moments in time: first, through the event represented; second, the moment of its conception and design; and finally, the moment(s) of its perception (Carrier 2005: 32). The Holocaust has been represented with two different collective memories in mind,

e 7.2 External s of Topography rror (left and e) and Holocaust orial (right).

and as such, their design and eventual perception should also differ. What is relevant in both cases is the question of cultural transmission, and the power to faithfully portray the intentions of the agency that created them (Jordan 2006: 60). It has become apparent that the main obstacle in the full understanding of a monument is actually its durability: by surviving their original contexts, memorials, and the built environment in general, have a tendency to lose their full sense, maintaining a durable material form without a clear, legible meaning (Jordan 2006: 60). By analysing the experiences gained during visits to these sites, the author aims to understand if the transmission of the memorials' messages are still readable, or whether, as Musil claimed in 1927, 'there is nothing in this world as invisible as a monument, [. . . since] anything that endures over time, sacrifices its ability to make an impression' (2006: unpaginated).

The design of the documentation centre at the Topography of Terror is fundamentally minimalist: a simple square building organized around an inner yard which features a pond in the centre. The building is entirely enclosed with glass walls and metal beams on the façade that play with transparency and opaqueness, light and shadow. The rest of the site has been carefully organized: the main outdoors exhibition areas are placed by the Wall and the edges of the site; paths on the coarse gravel have been created in order to guide the visitors towards the documentation centre. These inflexible and tidy pathways force the visitor to follow a determined circulation, to the point that the far end towards Stresemannstrasse seems to be out of reach: all of a sudden the path is not so well defined; the vegetation grows freely, out of control; signposts are non-existent even though there are remains of buildings. The authenticity of this corner is very strong, but is not properly acknowledged. The outdoor exhibition by the Wall has also been organized in such a way that there is no room for seeing the ruins without engaging with the now permanent exhibition: there is literally no space to follow the path without paying attention to the exhibition. From above, along the length of the Wall, the view of the ruin is also partially obscured by glass ceilings and exhibition panels. The Topography of Terror has been transformed into a topography of control, of tidiness that does not connect with the site, with the authenticity of the location, with the weight that the past and collective memory should have in that particular place. The documentation centre could have been located anywhere in Berlin: there is no anchor with the site; the circulation is not encouraging the visitor to look outside and reflect on the fact that some of the events narrated in the exhibition did actually take place there. There is no sense of 'terror' any more, just another kind of control over the communication with the audience.

The Holocaust Memorial was also controversial. A competition was opened in 1992 – the site being the former no-man's land, opposite the Tiergarten Park – and 528 designs were submitted. In 1995 the jury chose two projects, but they had to be disregarded as both received heavy criticism from the press and the public. A second competition was organized in 1997, aiming at including as many opinions as possible through several colloquia on the subject; the best entries from the first competition were asked to improve their designs by including the topics debated in these fora. James Young participated in the selection process and was concerned that such a memorial would '*unshoulder* [Germany's] memorial burden', and, once finished, memory would also finish (Young 2000: 194). Peter Eisenman's minimalist interpretation was the eventual winning proposal following many more years of controversy and public debate.

At both these memorials selection has played an important role in the representation of diverse and conflicting memories. It could be argued that the overarching aim behind most of Berlin's memorials is to include a multidirectional approach to memory, aiming to uncover overlaps and conflicts archived in the realm of memory and politics: justice can only take place when acknowledging all parallel memories (Rothberg 2009). The Topography of Terror represents the Holocaust through a largely black-and-white narrative: only perpetrators and victims are showcased – bystanders or German people who helped the Jews have not been acknowledged (these narratives can be found, for instance, in the Otto Weidt Workshop for the Blind, or the Memorial of Silent Heroes, also in Berlin).

Something similar happens at the Memorial to the Murdered Jews of Europe; the name by which it is mostly known, the Holocaust Memorial, is incorrect, since it does not incorporate, in its remembrance, other groups that were also persecuted by the Nazis, namely homosexuals, and Sinti and Roma communities. These groups have their own sites for memorialization close to the Holocaust Memorial (Figure 7.3). The Memorial to the Homosexuals Persecuted Under the National Socialist Regime is a simple structure, similar in shape and material to the Holocaust one, with a small window which invites the visitor to watch a short film about how homosexuals are perceived nowadays – an overly simplistic treatment of the subject. But if this monument is disappointing, the Memorial to the Sinti and Roma of Europe Murdered under the National Socialist Regime is an utter farce: the decision to erect such a monument was agreed in 1992, but work commenced much later in 2008, only to be abandoned in 2010 (Torisson 2010: 74). In July 2012 there were posters on the fences criticizing the lack of interest in finishing this monument. The passivity in developing the memorial demonstrates

re 7.3 Memorial e Homosexuals ecuted Under National Socialist me (left) and norial to the Sinti Roma of Europe dered under the nal Socialist me (right).

the connection between architecture and the political act of remembering; the selection, or lack of, a particular collective memory influences the construction of identity.

Several authors have understood the building of memorials and monuments as a moral obligation, or even as an opportunity to 'come to terms with the past'. However, one could argue that, in the case of the two buildings, this has not been very successfully executed. At the Topography of Terror the narrative has lost its authenticity through its overly careful spatial organization: there is almost no space for interpreting the ruins, which have been clinically displayed and represented; the original topography of the site has been silenced, erased. The first time the author visited this location (2006), the documentation centre had not been built and only the temporary exhibition was available as a source of information. The most staggering characteristic of the site then was the powerful simplicity of the void left in the city, and the few traceable remains of the building, which enabled the viewer to make connections between the building and the events that once took place. A strong presence of authenticity made the experience highly emotional, which connected to the concept of prosthetic memory, 'an interface between a person and a historical narrative about the past at an experiential site' (Landsberg 2004: 2).

These experiential sites encourage visitors to embrace the past in a more personal and emotional way. Furthermore, the remaking of significant episodes by different means of communication and their repetition seems to help societies to make sense of traumatic events (Sturken 2007: 29). The Holocaust or the terror attacks that took place in New York on 11 September 2001 serve as examples of episodes that, having been portrayed in different media, we all share a more emotional connection and understanding of them. The same cannot be argued, for instance, about Apartheid, the partition of India in 1947, or the terror attacks in Madrid on 11 March 2004. It seems the Topography of Terror is not really encouraging the potential empathy of prosthetic memory; the Wall, which describes an entirely different episode, has a stronger presence on site as an authentic remnant of the past. This lack of empathy with the site has a clear impact on the tourists of history, who aim to experience the past by consuming image, souvenirs, popular culture, museums and memorials in order to achieve a cathartic or more authentic experience of history (Sturken 2007: 9). As a tourist of history, the first experience engaged with the authenticity of the site and the aura of its ruins; it promoted a prosthetic memory and experience, which remained inert around the pristine new documentation centre.

The Holocaust Memorial, on the other hand, offers a public space which encourages social interaction in an unusual and even playful manner: we could observe people jumping between the pillars, seating on the lower ones sunbathing or simply resting; playing hide and seek and whistling in order to find each other. People were loud and did not necessarily read this minimalist structure as a place for reflection and remembrance. Furthermore, its location also promotes this attitude since it is in the heart of the Berlin tourist trail – between Potsdamer Platz, Brandenburg Gate and the Reichstag – just as another structure that can be read as architecture of spectacle, 'architecture as entertainment' (Smith 2006: 26). It seems as if the architect's and commissioner's interpretations of the historical episode had been overruled by the experience of the new generations of visitors, emphasizing Carrier's analysis of monuments. No matter the original agenda behind the erection of these memorials, it is impossible to control the audience.

The underground Documentation Centre at the Holocaust Memorial (Figure 7.4) offers a completely different interaction. We could appreciate how people behaved in a contrasting manner, as if the norms of conduct of a museum (documentation centre in this case, but very similar in essence) encouraged people to be quiet, paying attention to the information displayed. Light, circulation, sound and materials all promoted a more stimulating and personal experience, closer to prosthetic memory. The information revealed was clearly familiar to all the visitors, and as such the cathartic essence of the tourist of history's aim was easily achieved. Even though the Topography of Terror has been located on an authentic site, the Holocaust Memorial's documentation centre seems to be more powerful thanks to a design which aimed to enhance the visitor's experiences. The neutrality that the Topography of Terror is advocating contravenes a more personal approach to the site whilst inauthenticity and a more theatrical approach of the Holocaust memorial offer a better platform to reinforce prosthetic memory and engagement with the past.

As pointed out earlier on, Carrier analysed memorials according to three different moments: the episode that is represented, the moment of their design and erection, and finally, the different times of their perception. When we use these elements to analyse the Holocaust Memorial and the Topography of Terror, it seems quite clear that the symbolism behind these constructions has been transformed dramatically over time; moreover, it continues to change together with the new generations. Both monuments are in charge of representing the same episode, but, as explained earlier on, they have shown different collective memories in their structures. Both monuments were entangled in the debates surrounding the necessity of commemorating the Holocaust, especially since the 1980s; the generation of the immediate post-war which was dealing with

re 7.4 Internal 's of Topography rror (left) and caust Memorial t).

reconstruction, whilst the second generation requested the acknowledgment and commemoration of past wrongs (Fullbrook 1999). The 1980s also provoked the historians' conflict which debated how relevant the Third Reich should be in the history of Germany, and ultimately in the German museums of history (Maier 1997).

We are now immersed in the time of the perception of the finalized memorials, and it could be argued, following Halbwachs, that the main actor in the process is the public, the audience that is interested in remembering or forgetting, in acknowledging actively a difficult past, or blending it into the landscape of everyday life which encourages oblivion. Musil, Young and Carrier had already warned against this kind of memorialization: monuments are invisible, because time divests them of their context, of their meaning. The new generations are aware of the Holocaust thanks to a number of diverse sources: the topic is still quite fundamental in German education. However, the inheritance of the guilt (Schlink 2010) or shame (Dresler-Hawke and Liu 2006) from the past can only have a certain effect in the third and fourth generations of Germans.

Other groups may choose to commemorate the Holocaust in a more active way, depending on how relevant this episode is for their identity. For most tourists of history, the Holocaust is still rather recent in our collective and prosthetic memory; for the victims of the Holocaust, this episode is still at the core of their identities and as such, their approach to these memorial sites should be rather different. What is still to be learned is how the future generations will react when visiting and experiencing these sites: will the authenticity of the former Gestapo Headquarters be more important than the experiential architecture of the Holocaust Memorial?

After several visits to both sites between 2006 and 2012, the author believes that the Topography of Terror, an authentic trace in the built environment that has become a monument (unintentional, though), has lost the essence of its strength as spontaneous shrine (digging as an exercise of action memorialization): its authentic nature has been diluted in the process of institutionalization of memory. The attention has been shifted from the overwhelming scale of an immense void in the city centre of Berlin to a tidy organization that simply aims to inform. The indoors exhibition talks about facts in a very neutral fashion; the outdoors exhibition, even though it has been framed by the ruins of the former Gestapo Headquarters, is also offering just facts, whilst the panels act as a barrier between the audience and the authentic remnant, obscuring the different layers of history. The ruin has become heritage, and as such must be kept at a distance. The rest of the site, covered with coarse gravel, is rather unwelcoming: it makes it very difficult to walk on, giving the impression that wandering around the site is implicitly forbidden.

The Holocaust Memorial, an intentional monument, seems to offer more than one experience: remembrance and recreation. Between 2006 and 2012 the author was able to visit the Holocaust Memorial on several occasions and under different climate conditions. Rain can provoke a landscape of desolation: the 2,711 concrete slabs shine due to the anti-graffiti chemical that protects them, multiplying the effect of their grey colour. Moreover, the weather does not welcome many visitors, and as a consequence the place can be literally empty. On a sunny day, the Holocaust Memorial can be quite crowded: people wondering around the slabs, jumping from one slab to the next, seating on the lower pillars (which act as benches under the trees), or queuing in order to enter the documentation centre.

Whilst the first visit to the Holocaust Memorial was very emotional and the author empathized with the space – and the episode it commemorates – almost as if it were a shrine, a relic at the end of an emotional pilgrimage, the other visits showed the author a more convivial experience, with an intentional monument that has been interpreted by the audience into a playful and vivid public space. Even though the rigidity of the grid plan does not encourage interaction, it provides an interesting space to get lost in, to experience different views of Berlin, but also a space for reflection. In order to reinforce its function, there are several plaques that dictate how one should (or should not) behave in such a space, which includes not having barbecues, sunbathing in bikinis, and, of course, jumping from one slab to the next.

It seems that it is only in the context of the documentation centre that the audience behave in a more restrained manner. The question of behaviour begs a more relevant question: is it necessary to remember the past in a solemn way? Why is it so disrespectful to remember the past in a more informal way? Young noted that the first memorial project on the former Gestapo Headquarters was dismissed because it included recreational space; the current barren site seems to purposely neglect such human interaction, and yet it is the audience who should determine what and how to remember. It is interesting that in these particular two cases in Berlin the recreational aspect seems to be completely inadequate, whilst in other contexts, like the Berlin Wall Trail (2002–2006), Parque de la Memoria (2007) in Buenos Aires (Argentina), or the National Memorial Arboretum (2009) in Alrewas (United Kingdom) it seems to be at the core of their designs.

Conclusion

The aim of this chapter was to explore the relationships between memory and the built environment. The first section established the debates that surround the perception and construction of collective memories and identities, the selection of the past and its extension into the present thanks to the concept of prosthetic memory. The second section explored how the built environment can be read as a repository of memory, and how often it becomes invisible as it witnesses how time goes by. The third section analysed two particular case studies: the Topography of Terror and the Holocaust Memorial, both in Berlin, both commemorating the Holocaust using different narratives and collective memories. The main difference between both sites is actually the value of authenticity versus the impact of a more theatrical environment in order to achieve a better response as recipients of memory. After analysing the case studies against the theories on memory and

Figure 7.5
Trace of the Berlin Wall (left); Memoria in Bebelplatz, designed by Micha Ullmann (1994) (centre); Stumbling stones in Rosenthal Straße, designed by Gunter Demnig (199 (right).

field notes, it could be argued that both structures are successful as instruments of knowledge, keeping the memories in their documentation centres. However, their role in the encouragement of prosthetic (emotional) memory seems to be more limited.

Cubitt argues that cultural devices can have two very distinctive usages as mnemonic devices. First, they can aid in the storage of data; second, they can serve to memorialize events or people, with a clear moral imperative in its core (2007: 144–145). Both the Topography of Terror and the Holocaust Memorial have been designed with these two elements in mind: storing information and memorializing events and people. However, is this enough to reinforce the moral duty of encouraging remembrance? Perhaps it is necessary to combine these memory artefacts with a particular date or anniversary in order to reinforce their messages (i.e. The Holocaust on 27 January). It could be even more effective to introduce small memory devices that are not imposed on the landscape, that offer an invitation to learn more about the past. Less grandiose interventions (Figure 7.5) are more likely to catch our interest when we find them in the context of our daily – and why not recreational? – activities. By doing so we may become more aware of the past and keep debating whether the built environment can be read (and experienced) as a palimpsest; whether monuments are invisible or not.

Chapter 8

Identities in the architecture of Mexico City

Memories and histories in La Plaza de la Constitución and the Temple of Santa Teresa la Antigua[1]

Guillermo Garma Montiel

Introduction

Architecture and the city form the stage where our lives unfold; they frame our experience of the world and define our actions collectively and as individuals. They are more than a physical framework destined for human inhabitation; they embody the values and ideologies of the cultures that design, construct and occupy them. Buildings, in particular, are considered cultural artefacts that personify the values and ideologies of the society that produce them. As buildings are conceived, occupied, modified and in some instances abandoned, they become a complex ensemble of meanings, memories and values, which form a distinct sense of identity and place. These fragile elements and processes are constantly redefined and transformed despite having a definite physical presence in the world. By reading the city and its architecture it is possible to examine the meanings, memories and values intertwined in their material form and to understand the role these have in the construction of our sense of identity and belonging.

Joseph Rykwert described the modern city as malleable and in constant transformation; it is a place of contradictions, which 'houses many *ethnes*, many cultures, and classes, many religions. This modern city is too fragmentary, too full of contrast and strife' (2004: 7). Rykwert's words describe modern cities across the globe, from well-established to emerging metropolises, thus his words suitably characterize Mexico City's urban condition. The country's capital grew unevenly in the second half of the twentieth century, producing a metropolis where the old and new co-exist in close proximity, creating a surreal imagery of the capital. At the beginning of the twenty-first century, Mexico City epitomizes the tension between social and spatial orders; it contains within its amorphous boundaries stark social and economic contrasts, compounded by serious infrastructure deficiencies and pre-modern living conditions. The capital shows the benefits and disadvantage of modernization and modernity (Burdett and Sudjic 2007: 11, 170–171). This chapter focuses on Mexico City, and discusses the connections between the country's history and the memories and ideals embodied in its architecture by using two examples – La Plaza de la Constitución and the Temple of Santa Teresa la Antigua. I will argue that the meanings and memories of the past are tangible in the present, creating a fragmentary identity, which uses architecture to adjust to the demands of the globalized capitalist world.

Memory and identity

Architecture and the city constitute the stage where our everyday lives take place collectively and as individuals; they define and structure our existence in tangible and intangible manners. As Kim Dovey argues, 'architecture and urban design "frames" space, both literally and discursively' (2008: 1). In a literal or physical sense, architecture and urban spaces determine our actions, activities and experiences of the world; in a discursive and less tangible way they represent the processes, meanings and memories of the cultures that produced them. Architecture and the city are the physical embodiment of the historical past, representing not just the social, political and cultural endeavours of the societies that construct them, but 'they actually constitute much of [their] social and cultural existence' (King 1990: 404). It is in, and through, the built environment that societies and cultures thrive, engendering complex relationships embedded not only in built forms, but in objects, traditions, rituals and ideologies. Architecture and the city constitute and represent society and culture.

In the current globalized world, the postmodern condition characterized by fragmentation and multiplicity of meaning defines these relationships. In the last quarter of the twentieth century architecture and urbanism moved beyond modernist metanarratives, becoming an ensemble of traces from the past, ideas from the present and visions of the future. As David Harvey affirms, 'postmodernism cultivates [. . .] a conception of the urban fabric as necessarily fragmented, a "palimpsest" of past forms superimposed upon each other, and a "collage" of current uses, many of which may be ephemeral' (2007: 66). Architecture and the city are the result of a multiplicity of political and economic agendas, social and cultural values, and theoretical discourses, all of these compounded by the processes of globalization.[2] They document society's dreams and identities, histories and memories, and differences and struggles, all of them expressed through design, materiality, functionality, aesthetic proposals and intellectual discourse. The built environment, in all its conceptualizations, is a cultural manifestation that bears significance and contributes to the construction of place and identity. Today architecture's referent is the modernist / postmodernist debate of the twentieth century; it is there, with the revalorization of the past and tradition in the 1970s, where the ubiquitous preoccupation and impulse for preserving the self – identity – re-emerged. As Madan Sarup states: 'the past is the foundation of individual and collective identity; objects from the past are the source of significance as cultural symbols' (1996: 97); thus the past anchors our sense of belonging and strengthens our identity. Traditionally, identity was considered a coherent, fixed and unified phenomenon defined by social and cultural dynamics (gender, class, race, religion); today, however, it should be defined as a process in constant construction and fabrication, always in the act of becoming and replete with contradiction and ambiguities. Despite the wider impact of globalization on society, and the changes to global–local relationships, identity is related to the particularities of place and time. The self has numerous influences from across the world, but it is manifested in a particular place, living at a specific time and forming part of a determined culture. It is important to establish the nature of place, as there are several ways to conceptualize it. Place is a complex concept, and irrespective of the perspective that one adopts, it denotes a particular understanding of the world (Cresswell 2006: 79). Thus place, following Doreen Massey's ideas, is dynamic and is in a constant process of construction and definition. Its boundaries are porous, allowing slippages, influences and connections to occur, hence modifying inside–outside and global–local relationships. As place is not static, it is the site for multiple identities, histories and memories, which is formed by the social interactions that happen in it (Massey 1994: 155–156).

The construction of identity occurs in the present based on acts, events and recollections of the past; the stories that contribute to the formation of individual and collective identities are often about our past and dependent on our recollection of the events. It is in the act of remembering the past that memories and stories are selected, emphasized, transformed and erased, reconstructing a specific image of the past in the present. As Andreas Huyssen postulates, 'every act of memory carries a dimension of betrayal, forgetting, and absence' (2003: 3–4). It is this process of selection that makes memories malleable and elusive, as they are inserted, connected and determined by other events from the past. The reconstruction of the past can be controlled and manipulated, but it will always contain an unintended slippage of ideas, memories and meanings. According to Maurice Halbwachs, 'it is in society that people normally acquire their memories. It is also in society that they recall, recognize, and localize their memories' (1992: 38). For him, memories are connected with previous experiences that are reconstructed under societal pressure, and are manipulated in accordance with the conditions of the epoch. In his work, *On Collective Memory*, he distinguished between individual and collective memory. The former is constructed by the experiences of each individual and is bound to place and time; the latter is the memory emerging from the group, yet it is composed by individual memory. Collective memory serves as a framework for remembering, a phenomenon that is always taking place in society (1992: 51–53). There are a number of challenges to the concept of collective memory that highlight its contested nature. To start with, there can be as many collective memories as social groups, differentiations and pressures; all of these memories can be located in many places, having many manifestations, for example, myths, language, traditions, rituals, popular culture, archives and the built environment. However, the most serious challenge to collective memory when discussing identity is its selective nature, which prioritizes certain objects, memories, symbols and discourses over others in order to reconstruct a particular view of the past (Boyer 2012: 327–329). All these criticisms are valid, but collective memory remains an important tool for understanding the past and studying the formation of identity in the present. By being anchored within a society, collective memory can account for the complex relationships found within a culture. Furthermore, by studying the unofficial histories and memories of the past we can produce an inclusive and comprehensive collective memory in the present.

In conjunction with memory, history plays a key role in the construction of the present; the two are mutually dependent and respond to the cultural narratives and frameworks of a particular epoch. History can be defined as the set of objective facts and information stored and documented in written records, photographic evidence, material artefacts and the built environment; however, despite history's role in studying and learning about the past, its objectivity should be questioned, as it is articulated through the social, political and cultural agency of the historian. On the other hand, memory has a subjective dimension that relies on lived experiences; it is alive and actual, and contributes to the shaping of identity through time. Borrowing Enzo Traverso's words: 'History and memory are not opposites [. . .], but held in dialectic tension. The sources of history are constantly enlarged by memory: contesting its blind spots, bringing silenced and subjugated voices into play, replacing amnesia with remembrance and reconciliation' (Traverso cited in Boyer 2012: 329). One of these sources is the built environment, which embodies the material traces of the historical past, becoming a meaningful reminder of its significance and importance. Our sense of reality emerges from being *in* the world and inhabiting the spaces of the built environment; as Maurice Merleau-Ponty affirms, 'I am a body which rises towards the world' (2002: 87) and I perceive it through the corporeal inhabitation of its spaces. Architecture and the city provide the built forms for the sensuous experience of the world, which in turn creates the lived experiences that form our individual and collective memories.

Architecture and the city are mnemonic artefacts formed by a variety of ideas and meanings, which are constantly overlapping, transforming and interweaving, depending on the historical moment and the cultural, social and political groups that validate and appropriate them. Thereby architecture's original intent and meanings are subject to appropriation, reinterpretation and development. To read architecture as a cultural artefact is to unveil the forces and processes that influenced it, examining its anatomy in search of cultural insights (Sharr 2012: 2–3) and conceptualizing it as a dynamic object representing diverse identities and communities. These ideas will guide the analysis of the historical centre of

Mexico City, La Plaza de la Constitución and the Temple of Santa Teresa la Antigua, which I will use in order to examine the construction of identity in Mexico.

The masks of Mexico: celebrating history and memory in the Zócalo

Mexico was born in the sixteenth century from the confrontation of two great civilizations: the Spanish and the Aztec. Its cities – and architecture – are the expressions of a society produced by the processes of conquest, colonization and syncretism that took place after the fall of Tenochtitlán in 1521. Even though the Spanish imposed their worldview through the sword, language and religion, eventually they reached a level of syncretism with the indigenous cultures. Carlos Véjar Pérez-Rubio explains that 'one factor characterizes the Mexican environmental culture: syncretism, the fusion of ideas, beliefs, likes, experiences, construction materials, techniques and diverse life styles' (2007: 67). The fusion of the pre-Hispanic and Spanish cultures is illustrated by the foundation of Mexico City. The first urban act of Hernán Cortés in 1521 was to order Alonso García Bravo to outline and start constructing the new capital of the empire over the remains of Tenochtitlán. García Bravo outlined the city based on classical European urban concepts (two main avenues, perpendicular to each other, originating from the Royal Square) and used the existing infrastructure and the vestiges of the main Aztec structures to anchor the new Spanish buildings. Despite the geographical disadvantages of the small island where Tenochtitlán was founded, his decision responded to strategic advantages, but most importantly to ideological and symbolic motivations. Cortés wanted to erase the Aztecs' religious and political institutions by physically substituting them with the Spanish equivalents (Espinosa López 2007: 27). Hence the initial urban design of the Spanish city fundamentally respected the layout of Tenochtitlán and unwittingly produced a physical syncretism between the two civilizations. The adoption of Tenochtitlán's urban layout can be interpreted as the translation of the pre-Hispanic city into the capital of the Spanish empire in the New World (Figure 8.1). As Joseph Rykwert noted, 'the centre of Mexico City is urban in the old sense. It retains the features of a Spanish colonial town around the Zócalo, the main square, which Cortés laid out just after the Conquest' (2004: 7).

The Aztec's ceremonial quarter, located at the heart of Tenochtitlán, was demolished and new constructions were erected in it place. In the south, in the space between the palaces of the Aztec rulers Moctezuma and Axayacatl, the conquerors laid the city's main square. La Plaza Mayor, as it was known in the sixteenth century, became the locus of the political, economic and religious powers of the colony; it was surrounded by arcaded buildings for trade, the National Palace, the Cathedral, the town hall and the law courts, all of them embodying the physical decapitation of the indigenous system of beliefs and its replacement by a new historical discourse of discovery, salvation and colonization. The built environment was used to impose a discourse that excluded the natives from history and installed a European history devoid of indigenous memories, rituals and

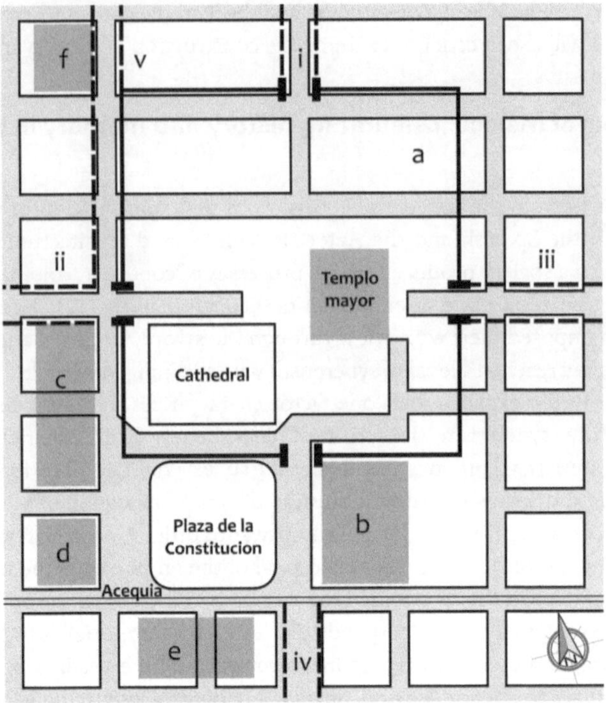

Figure 8.1 Plan showing the juxtaposition of the Aztec's infrastructu and main buildings over a plan of Mexico City: a. Aztec's ceremonial centre according to Marquina, 1961; b. Moctezuma's new Palace; c. Axayacatl's Palace Cihuacóatl's Palace e. Nobles' houses; Cuahtémoc's house Calzada to Tepeyac Calzada to Tlacopa iii. Towards Texcoc pier; iv. Calzada to Iztapalapa; v. Calza to Tlatelolco.

values (de Anda Alanís 2006: 75–77). For the next 300 years the Spanish controlled, shaped and exploited the colony's human and natural resources, generating in the process a hybrid national identity defined in relation to Spain. In the eighteenth century, the descendants of the Spanish and the indigenous people, or *criollos*, began to affirm their national identity based on three principles: connections with the territory they inhabited, the rescue of the indigenous past and the creation of new patriotic symbols (Florescano 2002: 269–270). The emergence of a national identity, the search for social equality and the desire for a political voice were contributing factors to the outbreak of the War of Independence in 1810.

In the early nineteenth century La Plaza Mayor was renamed as La Plaza de la Constitución in honour of the 1812 Cadiz Constitution,[3] but the square was commonly known as the Zócalo. This name originated from a planned monument to celebrate the War of Independence (1810–1821) that was never constructed beyond the base or *zócalo*. The north side of the square is delimited by the Cathedral, which symbolizes the new faith of the nation that was imposed on the indigenous population from 1521 onwards, subsequently becoming deeply rooted in the Mexican psyche and society. In 1978 the inheritance of the Aztec past was unearthed when the vestiges of the Aztec's Templo Mayor were discovered and exposed by order of the president José López Portillo (de Anda Alanís 2005: 182). The Templo Mayor brought to the surface the ancient faith of the country and contrasted it with the Catholic heritage, producing a juxtaposition

[Fig]ure 8.2 In the [for]eground we [ca]n appreciate the [Te]mplo Mayor's [ar]chaeological site, [in] the background the [Mu]seum of the Templo [Ma]yor, by Pedro [Ra]mírez Vázquez [an]d Jorge Ramírez [Ca]mpuzano (1987).

of memories and truths, both constructions embodying architecture's agency in moulding the cultural, religious and political relationships in the New Spain (Figure 8.2). La Plaza de la Constitución is a national symbol embedded in the collective memory of the country, acting as the locus of the country's political life and the hub of the capital. It represents the meanings, memories and ideas of the country, symbolizing the construction of a hybrid Mexican identity, where the indigenous past is merged with the Spanish inheritance and then exposed to contemporary ideals of modernity.

The Zócalo has framed historical events such as the proclamations of the King and the Viceroy during the viceroyalty, but also has validated the country's history by celebrating the anniversaries of the Independence and the Revolution. The memories of these historical events have shaped the collective memory of the country, becoming an anchor in the national consciousness by the symbolic importance of the square and its surrounding buildings. The architecture not only delimited the occurrences, but also drew from the reservoir of past memories embodied in the square to enrich the culture of the country. The rituals, ceremonies and public gatherings that take place in the Zócalo call attention to the Mexicans' inclination for fiestas, which serve as the perfect stage for displays of freedom and national pride to take place. In the words of Octavio Paz, 'the solitary Mexican loves fiestas and public gatherings. Any occasion for getting together will serve, any pretext to stop the flow of time and commemorate men and events with festivals and ceremonies' (1985: 47). It is in fiestas and other 'ceremonies – national or local, trade or family – the Mexican opens out. They all give him a chance to reveal himself and to converse with God, country, friends or relations' (Paz 1985: 49), strengthening his relationship with the national psyche. It is on these occasions when the Zócalo, and its architecture, becomes a place where Mexicans dwell and embrace their diverse inheritance. It is during these rituals, celebrations, rallies, protests and events that the square is appropriated and its meanings are transformed, opening the possibility for a new identity based on the memories of the past and the identities of the present. For any Mexican, the experience of celebrating the Independence in La Plaza de la Constitución and the adjacent streets is a vivid reminder of the juxtaposition of historical events and memories that contributed to the construction of the nation. The individual does not occupy abstract space, but by defining meaningful routes and activities in the square, one constructs the city that one inhabits, creating individual memories and a sense of place. The architecture of the historical centre structures a heterogeneous image of the capital, based on a range of sources and interpretations that may be unpredictable, imaginary and even opposed to official narratives. By reconstructing the urban image of the city centre with meaningful experiences of the present, along with the historical built forms and memories of the past, the city is absorbed in the national psyche and is incorporated in the inhabitant's memory. The cracks and discrepancies between the memories and histories of the past offer the opportunity for alternative recollections and insights to slip through to the unconsciousness, creating a sense of belonging and shaping the individual and national identity.

The Temple of Santa Teresa la Antigua: moving into modernity

After centuries of conflicts and transformations, the Zócalo emerged as an urban palimpsest embodying the physical and ideological confrontations of the pre-Hispanic and Spanish memories. It served as the locus of national identity and pride, where the past was celebrated whilst confronted with the realities of the country's present. Despite its complex nature, the subtle changes to the square's morphology[4] complicate the analysis of the conditions that influenced architecture in the last two centuries. Hence I will examine the Temple of Santa Teresa la Antigua in order to read the impact that post-Independence events had on architecture and the national identity. In colonial times, architecture was defined by a process of hybridization, especially in the construction of the main typology of the time: religious buildings. The early churches, temples and convents presented the opportunity for the syncretization of Spanish and indigenous cultures. These religious buildings were designed by the Spanish priests/architects using European canons of construction, but with local materials and indigenous labour force; hence the artistic spirit of the native craftsmen was expressed in the decoration of the facades and the internal spaces of the buildings (de Anda Alanís 2006: 78–86). The architecture of the colony mirrored, to some degree, the emergence of a hybrid national identity in the country.

The convent of Santa Teresa la Antigua was founded on 1 March 1616 in two dwellings 300 meters away from La Plaza Mayor. The two properties were donated by Don Juan Luis de Rivera and were refurbished in 1615 to accommodate the needs of the monastic order of the Carmelitas

Descalzas. Initially the convent's first church was hosted in rooms within the convent; in the middle of the seventeenth century, however, the construction of the current church was started, and it was completed and consecrated by September 1784. The expansion of the convent did not stop there and in December 1798 the architects Antonio Velázquez and Manuel Tolsá started work on an external baroque chapel, which was completed in May 1813. The chapel was named Capilla del señor de Santa Teresa, in honour of the religious image of the Señor de Santa Teresa, which it hosted as soon as it was completed. Further work was required when the chapel's apse and dome collapsed on 7 April 1845 due to an earthquake, resulting in the construction of a new apse and dome by the architect Lorenzo de la Hidalga (Rossell 1979: 287–290). The simple spaces and strict spatial planning of the convent relate to the religious order's asceticism and observance of their vows of poverty. As with most of the Carmelitas Descalzas' religious buildings, it was characterized by a strong use of unbroken horizontal lines, simply decorated surfaces and simple geometries in the design of its spaces. The religious complex typifies the monastic order's tendency to use the convent plan with narrow cloisters and an adjacent lateral chapel in their constructions; the life of the convent looks inwards, whilst the church and chapel serve as the point of contact with the congregation (de Anda Alanís 2006: 100–101). For Mass, the nuns were segregated from the congregation, entering the church's nave through the high and low choirs, which were located facing the main altar and on the opposite end of the presbytery. The need for seclusion determined that the nave ran parallel to the street and the public entered via two twin elevations leading into the middle of the nave (Figure 8.3).

In the two centuries since its foundation, guided by the order's ideals of austerity, the convent's religious community increased and became prosperous. However, this changed after 1857, as the troubled years that followed

re 8.3 Main ation for the ent of Santa sa la Antigua.

the movement of Independence (1810–1821) culminated in the War of Reform (1857–1861), which redefined the influence of religion on society. The civil war saw the Liberal and Conservative forces fighting over the political control of the country. Eventually, after three years of hostility, the Liberal party, led by Benito Juarez, emerged victorious in 1861. The impact on the Catholic Church and the military was significant, as both institutions were stripped of their traditional rights, powers and properties (Cosio Villegas *et al.* 2003: 108–118). With the expropriation of the properties of the Church, the convent of Santa Teresa la Antigua was stripped of its religious connotations and allocated other uses that would reinforce the national project. The convent was used for educational purposes, becoming the first School for Teachers, the School of Dentistry and a sixth form college. At its closure in 1863, the convent numbered 22 women, and it owned 26 farms that produced a constant income for the subsistence of the nuns. Despite the loss of the convent, the church remained active until it closed its doors to worship in 1930 (INBA 2009 online). In the last three decades of the nineteenth century, Mexico found stability and development in the presidency of Porfirio Díaz (1876–1880 and 1884–1911), who governed the country until the War of Revolution (1910–1920).

At the dawn of the twentieth century Mexico emerged from the War of Revolution into modernity. The expectations and uncertainties of the new nation faced a devastating reality: after years of conflicts, the country's infrastructure and built environment were reduced to rubble; more than a million people died during the Revolution and the newly formed government realized that the country lacked a unified national identity to face the future (Florescano 2002: 393–395). In 1929, after nearly a decade of internal conflicts between the rival factions of the Revolution, the Partido Nacional Revolucionario was founded. The official party consolidated the revolutionary groups, and its leaders, under one political ideology, and set to use culture and education to 'integrate the dispersed ethnic, social and cultural mosaic into a project of national unity' (Florescano 2002: 399). In the five decades that followed the War of Revolution (1920–1970) architecture was tightly connected with the different political agendas of each administration; hence the majority of the architectural boom between 1920 and 1970 happened in state-sponsored public architecture, such as social housing, schools, markets and hospitals, which aimed to elevate and strengthen the profile of the Revolutionary doctrine (de Anda Alanís 1998: 31–32).[5]

As a consequence, after 1930 the Temple of Santa Teresa la Antigua had a variety of functions and it was used as barracks, the headquarters of the presidency of the Universidad Nacional Autónoma de México and the Universidad Vasconcelos. Also it was the print shop for the Diario Oficial de la Federación –the government's official journal – as well as the archive for the Secretaría de Hacienda (INBA 2009 online). These different uses aimed to reinforce the 'new' national political project that had emerged after the War of Revolution. Thus the nineteenth and early twentieth centuries were characterized by social, religious and political conflicts, all of which had a profound impact on the temple's morphology, design, function and identity.

Modernity, identity and architecture: Ex Teresa Arte Actual

The Mexican government's desire to modernize the country and participate on the international stage eventually led to the adoption of the ideals of modernism and its technologies to provide the necessary infrastructure and urban spaces in the city. Immediately after the War of Revolution, the European eclecticism favoured by President Porfirio Diaz was replaced by neo-Colonial and neopre-Hispanic architecture. These two styles eventually gave way to the abstract language and orthogonal geometrical composition of the International Modern, known in Mexico as Functionalism (Mendez-Vigata 1997: 62–63).

> Both the neo-Colonial and neopre-Hispanic were overtly nationalistic and expressed the will to portray a national identity in opposition to foreign influences. Thus, it was an approach aimed at preserving 'traditional' Mexican culture and values. The International Modern was thought to project the modernness of the post-Revolutionary government and the hope of a new future, a future that would include Mexico among the most 'progressive' nations in the world.
>
> (Mendez-Vigata 1997: 61)

Suitably, architecture was used, up until the 1960s, to embody and develop the national identity, to fulfil the social needs of the country and to construct a modern future for Mexico. However, the last three decades of the twentieth century were characterized by increased globalization and political and economic instability, exemplified by the severe financial crises of the 1980s and also of 1994, which forced the government to adopt economic austerity measures resulting in withdrawal from its role as sponsor of public architecture (de Anda Alanís 2006: 236). This opened the door for multinational corporations to invest in infrastructure, producing private corporate buildings with little architectural value, resulting in a distancing from the social and cultural reality of the country. The majority of these companies used an imported architectural identity of high-rise buildings, modern materials and daring formal statements to convey a global image of power, prestige and exclusivity, disregarding a sense of human scale, contextual particularities and the country's architectural culture.

Accordingly, in the last quarter of twentieth century the Temple of Santa Teresa la Antigua was refurbished twice, first in 1978 by the Secretaria de Asentamientos Humanos y Obras Publicas and then in 1993 by the architect Luis Vicente Flores, who also built an extension to the convent to host INBA's (Instituto Nacional de Bellas Artes) cultural project of Ex Teresa Arte Actual (Ex Teresa).[6] Flores' intervention represents the most significant addition to the complex in the last century and it actualizes the building, bringing it into the twentieth century (Figure 8.4). Ex Teresa was founded in 1994 with the purpose of providing the conceptual platform from which the public could develop an understanding of the language of contemporary art and to accommodate objectless and unconventional art expressions, e.g., video, sound art, installation art and performance

Figure 8.4 Ex Teresa Arte Actual's extension, architect Luis Vicente Flores (1993).

art. It emerged amidst a number of alternative spaces created for unconventional art expressions in the 1980s and early 1990s (La Agencia, f. Adolfo Patiño, 1987; El Salón des Aztecas, f. Aldo Flores, 1988). These spaces were the consequence of the socio-political movements of the 1960s, especially the student movement of 1968, and the emergence of Mexican performance art during the 1970s. These spaces welcomed interdisciplinary artistic groups with a strong social conscience and who were concerned with the articulation of identities of minority, subaltern and dissidence; all of them were in search of new platforms on a national and international level (de la Garza 2008 online).

Ex Teresa's mandate and artistic programme transformed the temple into a dynamic place constantly reconfigured by the art installations and performances; furthermore, the boundaries of the buildings become permeable as they are inserted in a national and international network of cultural connections. Firstly, at an urban level, Ex Teresa is anchored by its own history and memory, which interacts with those of the historical city centre; its architectural spaces are actualized and enriched by the visitors, the public and the performing artists. Secondly, on a national scale, it connects with cities and art organizations across the country. And finally, Ex Teresa's events attract international artists, who strengthen its global connections, creating a productive cultural exchange. The artistic manifestations hosted by Ex Teresa re-energize its architecture, producing new memories, challenging established histories and giving it new identities. The sacred nature of the Temple of Santa Teresa la Antigua is questioned through contemporary secular means. For example, in August 2008, Ex Teresa hosted *Inoculación*, a monumental art installation/architectural intervention by Antonio O'Connell, which criticized society's nihilistic principles, consumerist identity and absence of spiritual values (Figure 8.5). The installation used the slippages between art and architecture to reflect on their nature and relationship.

In *Inoculación*, O'Connell argued that from a very early age we have been conditioned by the media and advertising to base our happiness on the acquisition of material possessions – consumerism. Therefore, the 20 meter long by 8 meter high intervention represented O'Connell's twofold reflection: firstly the need to move away from material values, towards a renovated relationship with God; and secondly, the importance of architecture as the concretization of a deep human spiritual need transcending the architect's ego and the primordial need for shelter (O'Connell 2009). The building, including Flores' extension, was invaded by a complex sequence of sculptural elements and architectural spaces that metaphorically inoculated the buildings against the virus of consumerism. *Inoculación* criticized modernism's approach to the past and inspired the debate between the identities of a nihilistic, individualistic and hedonistic society, and an ascetic, spiritual and communitarian view of the world.

re 8.5 A key ent of O'Connell's *ulación* was an ' to consumerism h was positioned e space reserved e old church's ytery.

For the refurbishment and extension of Ex Teresa, Luis Vicente Flores removed all the previous interventions and used prefabricated elements that were easily dismountable to cater for the museum's needs. He then allocated the offices and services in a steel and glass lightweight structure at the back of the building; this modern box was annexed to the existing building without intervening into its original structure (Adrià 1996: 38). It could be read as a removable extension that respects the history and features of the religious building, giving it independence by avoiding a formal confrontation with it. However, despite the many merits of Luis Vicente Flores' lightweight structure, its physical detachment embodies its distancing from a frank conversation with the building's past history and memories. The modern structure borrows from modernist ideals and ignores the convent's historical context and religious content (Figure 8.6). The removal of all past interventions reflects a '*tabula rasa*' approach to the building,

Figure 8.6 Detail Luis Vicente Flore intervention. The image shows the relationship betw the modern exter and the historic building.

depriving it of its recent history. It can be argued that the new function of the building demanded such removal; however, Flores produced a modern intervention, without any emotional engagement with the building's past, either recent or remote. Those lost meanings and memories were an intrinsic part of the building's identity and his proposal should have accounted for them. The modernist intervention becomes a 'mask' of modernity that is imposed on the convent, distracting from its past identities and memories. The intervention responds partially to the conditions surrounding it by respecting its historical value; however, it disconnects from the past by using an imported architectural language and ideology. Its transparency reinforces the notion that it is an empty canvas without a sense of history and with a weakened architectural proposal; in a sense it is a premonition for O'Connell's virus. As *Inoculación* vaccinated Ex Teresa against the virus of consumption, Flores, inadvertently, inoculated the building against modernist glass and steel architecture that disregards the context and history.

Final reflections

As the effects and consequences of globalization define the contemporary world, consequently the transformations triggered by it at the local, national and global levels force us to re-evaluate the role of the traditional markers of identity and place. Therefore, it is important that the built environment be approached and conceptualized in relationship with its broad context, past and present, and for us to be aware of the fragmentation that this juxtaposition may engender. Despite its physical presence, architecture becomes a cultural artefact in a constant process of change and with multiple interpretations. Architecture and the city embody the histories and memories of the societies and cultures that produced and appropriated them, becoming important sites where memories, histories and identities are constructed and consolidated. For example, the Zócalo embodies the syncretism of the Mexican culture that was born from the clash of the Spanish and Aztec civilizations, and which set the foundations for the national identity of the country; on the other hand, Ex Teresa expresses the changes that have operated since the sixteenth century, and reflects the twentieth century's realities of an unfinished project of modernity. In both cases their symbolic importance is transformed by the activities and events that occur there (national celebrations and artistic manifestations), changing their role in the national consciousness.

The negative effects of globalization were evident in Mexico's financial crises, which determined the national development of architecture at the end of the century, impacting Flores' intervention in Ex Teresa. However, the loss of identity in contemporary Mexican architecture is also the result of a weakening of the architectural discourse produced by the separation of practice and theory, and the adoption of a pragmatic instrumentalism predicated on economic and political gain, rather than founded on sensitive historical and theoretical understanding (Véjar Pérez-Rubio 2007: 10). As Alberto Pérez-Gómez affirms, Mexican architects in the nineteenth and twentieth centuries considered theory as 'applied science' and of direct consequence and applicability to practice, rather than as 'an explanation of how certain things make sense in the realm of practice' (Pérez-Gómez in Burian 1997: 22–24). The understanding of theory became instrumental and it signalled the desire for a direct applicability of the architectural discourse. At the start of the twentieth century, and after centuries of cultural dependence and imposition from Europe, architecture relied on the reproduction of foreign models and solutions, especially those coming from the Beaux Arts academies in Paris, London and Rome. The eventual adoption of Functionalist ideas towards the middle of the century brought a strong influence of Le Corbusier; however, this was compounded by a renewed look at the country's past and an attempt at its reinterpretation. This illustrates an underlying tendency, where validation and recognition in the domestic context depend on imported principles and ideas from the global centres of power, fostering what Véjar Pérez-Rubio calls an 'environmental neocolonialism' (Véjar Pérez-Rubio 2007: 10).[7]

It is by reading architecture closely – as demonstrated through the two case studies – that the mechanisms, ideologies and questions that originally influenced them are revealed, allowing us to access the histories and memories that contribute to the construction and transformation of identity from one epoch to another. Architecture, as identity, is always in a constant process of becoming. The Temple of Santa Teresa la Antigua and the Zócalo constitute and reflect a variety of identities and memories, some complementing each other, others in frank opposition, but all of them finding a place in the hectic urban fabric of the centre of Mexico City. It is through the memories and histories embedded in these buildings that we find an entry point to understand the construction of Mexico's national identity in the twentieth century.

Chapter 9

Post-conflict space

Mediating memory and the parameters of identity

Fiona MacLaren

Introduction

Post-conflict space forms a particular and complex topography of place. A mnemonic landscape potently charged with divergent cultural, social, political and economic agendas, the relationship to space is experienced within physical, emotional and psychological parameters of place and identity. Spatial transformation is shaped by initiatives that strive to imagine, create and represent place beyond conflict. Urban regeneration, memorialization, cultural investment and historical representation are the contending factors in the physical re-shaping of the environment. In the discourse of directing change, these vying factors shape an altered physical and psychological landscape, able to preserve or erase spatial landmarks of history. The past is an integral component in the negotiation and determination of the future, and the extent to which historical representation is manifest in spatial form reveals the complexity of the post-conflict condition where essentialized versions of history belie the diversity of experience and multiplicity of social voice. Reflected in the instrumentality of memory, the dualism of remembering and forgetting interweaves the problematic engagement of a plurality of historical experience. As the events of conflict and trauma give way to a scarred and altered terrain, memory intersects the relationship of identity and place, whereby the dichotomy of remembering and forgetting becomes spatially manifest. Claimed and contested, memory is both barrier and point of access to a greater interpretation and understanding of the past, the processes of representation and memorialization that negotiate the meaning of the past to the future beyond conflict. Informed by theories within cultural geography, anthropology, architecture and visual culture, this chapter examines the triadic relationship of identity, place and representation to consider the concept of memory as a connective yet problematic medium that extends the reading and configuration of the post-conflict environment.

Memory: agency and duality

In recent decades, memory has been centrally positioned in the cultural field offering a potent methodology to critique the representation of the past. Brought to a critical reading of history, liberation, recognition and representation are positive underpinning principles. The agency of memory can reveal the plurality of voice and divergence of memories, individual and collective, that counter and problematize dominant, official or singular narratives of the past. However, the turn to memory, evidenced spatially in the proliferation of museum and memorial building, is aligned with the acceleration of history and reflects a deeper social anxiety in the face of the rapidly shifting technological conditions of cultures in the contemporary developed world.

The work of Pierre Nora indicates that an excessive focus on memory has globally established close ties between respect for the past – whether real or imaginary – and the sense of belonging, collective consciousness, individual self-awareness, memory and identity (Nora 2002: 1). The conditions of mobility, speed, dislocation and complexified temporal–spatial relations undermine stability and generate a state of flux that further compounds a fixation with memory. In what Nora describes as the age of commemoration, the crisis in the ideology of modernism and the perceived threat to memory from the post-modern condition has destabilized our sense of continuity.[1] The relational rupture between history, memory and identity bears parallel association to a contemporary cultural fear and anxiety harbored towards the future. Traditionally, meaning in the present was in part determined through the assured knowledge of what was required of the past to prepare for the future. Unable to distinguish what the future requires of us, or indeed what form it will take, we are seemingly compelled to preserve everything, archiving, storing, collating and stockpiling against the anxiety and inevitability of forgetting.

Remembering is central to individual and collective well-being, but infinite preservation is impossibility. The pursuit of ultimate preservation transforms memory into a burden and counters the aim it seeks to achieve; forced remembering can be seen in parallel to a need of controlling and censoring the past. Remembrance practices and memorials instigated out of obligation and the duty to remember diminish contextual interpretation and a critical understanding of history. Adam Phillips examines these concerns from a psychoanalytical context and reinforces the point that an obsession with memory blinds us to the abuses of memory and to the uses of forgetting (Phillips 2007: 134). Like memory, the past is partial, subjective, fragmented and fallible, open to contextual manipulation and appropriation. Within the cultural focus on memory and memorialization lies the perception that memory sits wholly in tandem with individual and collective wellbeing. Forgetting implies failing in the duty to remember and the abandonment of memory. Phillips explains our fear of forgetting resides in the anxiety over remembering correctly, and, in the context of trauma, we are unable to tolerate ourselves as beings that could forget atrocities. Reminding ourselves of our capability for atrocity is perceived as perhaps a false defense of their repetition. Phillips states: 'we equate forgetting with the trivial or the unbearable [. . .] *we have a place to put the unbearable* [. . .] but we believe that, the memory, the experience, the desire, are still there . . . capable of returning' (Phillips 2007: 134). In considering what it is that we want from memory, Phillips presents us with dichotomy that what we ultimately want is a *way* of remembering that will guarantee the impossibility of recurrence (Phillips 2007: 134).

In spatial terms, the polarity of total erasure and ultimate preservation mediates the extent to which the events of history configure the post-conflict landscape. If retained by expansive memorialization society risks being bound under a duty to remember, anchoring the present society under obligation to the past. Preservation in excess transforms place into numinous space, the social environment into a form of museum. It is widely discerned that remembrance requires material loci; sites of memory pepper post-conflict space. Often contested, they evidence the political instrumentality of memory whereby non-consensual interests instigate the struggle over representation; what do the sites mean and to whom? What and whose memories are evoked? While the contestation of sites of memory can retain them as sites of conflict, the disputed process of interpretation forces the respective debates for preservation, erasure or construction and as such acts as a catalyst of spatial transformation. In the dialectic of remembering and forgetting, the loci of memory is crucial to the practices of remembrance but they must negotiate the fragile balance between the double-bind of our want and need of memory, and the permission and necessity of forgetting.

Northern Ireland: memory, place, identity

A sense of place emerges through the interplay of memory and myth, politics and power, and the narratives of history. Like memory, the meaning of place does not inherently exist; rather it is produced through the representation of the past in the present and the contextual shifts in the experience of everyday life. Memory becomes inscribed in the landscape through the materiality of place. Under conflict the sense of place is transformed into a territorialized geography of spatial limitation; defenses, perimeters and boundaries of inclusion and exclusion. Rupturing the normality and security of everyday space, conflict re-frames the social, physical and psychological relationships to place, destabilizing the parameters of identity and conception of belonging. It must be stated: it is not being suggested that the contextual specificity of conflict can be conflated into a singular universal definition, however, a level of commonality is reflected through the central role the experiential affect of violence plays in post-conflict spatial transformation. For Northern Ireland, the mediation of memory and space manifests a complex interweave of the politics of place and identity. The violent and traumatic events of the period known as the Troubles sit well within living memory. Non-consensual interests reflected in political, social and cultural divisions frame the problematic of resolving shared space. Here spatial transformation is challenged by the co-existence of two differing relationships to place within opposing constructions of the same history.[2]

Belfast has undergone extensive development in recent years and expansive spatial development continues. It hosts all the attributes of a contemporary global metropolis; acclaimed restaurants, leisure and entertainment venues, luxury hotels, museums, galleries and an array of shopping facilities.

The representation of a city beyond sectarian conflict sat in parallel to the necessity of economic recovery and the regenerative means to readily compete in the global market of consumer capitalism. While such change was essential and supports the process of normalization, the presence of successful commercial enterprise does not efface the politics of identity nor erase the reality of social and spatial division. To this end, the rendering of Belfast's city centre as a neutral space for its inhabitants was met with considerable critique. Under the commonly used strategy in urban regeneration and the promotion of heritage, Belfast's centre was fashioned into quarters: Titanic, Linen, Cathedral, Queen's, Gaeiltacht, Library and Market. Re-presented in a framework of straightforward historical and cultural reference, the space is easily navigated promoting landmarked sites of tourist interest from the industrial past of manufacturing and shipbuilding to the attractions of a vibrant contemporary arts and performance culture.

Writer and academic Aaron Kelly commented of the risk in the substitution of sectarian division with new forms of segregation, in effect pointing to the increasingly polarized division of wealth through the forces of global free market capitalism. To appropriate culture as a 'byword for consumerism was to repress the political by re-presenting the cultural quarters of the city back to its people as spaces of consumption eviscerated of their experiences of it' (Kelly 2005: 548). The notion of a neutral area of the city intends to foster mutually inclusive shared space that functions without the boundaries of division. Yet by definition neutrality prevents contention and so precludes a more critical response to spatial representation that can engage the complexity of history and authenticity of place. Referring to a lack of unifying narrative of place, the concept of placelessness, in opposition to a sense of place, reinforces community division of Northern Ireland.[3] Cultural geographer Brian Graham explains 'the protestant community in the North has no confident text of place provided by a representative landscape' (Graham 1994: 266 quoted in Duffy 1997: 77) yet the text of place within the Catholic community is strong and secure. To this end, the divergent relationship to place fundamentally problematizes how authentic narratives of place identity and memory may be inscribed in the spatial shifts of post-conflict transformation, which occur in parallel to consumer capitalism; the condition of global urban space within which the pursuit of economic growth takes precedence. In the re-imagining of post-conflict space, memory, place and identity are potentially transformative agents in themselves, yet here, they also serve to sustain the parameters of historically bound identity and the spatial boundaries of inclusion and exclusion. In order to address the possibility of an authentic narrative of place, it is necessary to consider how, and by what methods, critical engagement with space and place may be realized.

Visual and spatial practice

Critical interpretation of the past is not to construct and sustain an infinite state of un-resolve. Rather by critically attending the representation of history, plural narratives and multiplicity of meaning may be permitted. Brought to the politics of identity, the visual and spatial practices of art and architecture can make critical spatial readings and responses that both problematize and reveal the intimate cultural and historical specificity of place. Recent writing by Vikki Bell examines the relationship between art and the processes of conflict transition and draws out the capacity for critical intervention inherent in the methodologies and practices of art. Framed through Benjamin's notion of constellations, she analyses how artworks can produce critical constellations that 'engage the cultural processes of transition through their problematization of it' (Bell 2011: 5). She expands that 'artworks by definition remain open to reinterpretation and new understandings [and as such] become sites in which the assumptions of transition are opened up for critical reflection probing the notion of what constitutes peace and its conditions of possibility' (Bell 2011: 5). In this way art and spatial practice is arguably pivotal to the processes of post-conflict transition whereby the capacity to critically engage in the politics of place, memorialization and the representation of history extends understanding and interpretation of the past's meaning, relevance and influence to the present.

This can readily be seen in work undertaken at Paragon Studios / Project Space known as PS², a small artists' collective established in 1993 that draws upon a breadth of practitioners and theorists in response to urban intervention and social interaction. In 2006 they launched the Space Shuttle project, a series of six 'missions' carried out by different artists within various locations across Belfast. The structure of the Space Shuttle took the form of a portacabin, fully mobile and scaled as a replicated model of the PS² project space in Belfast. The work of artist Aisling O'Beirn engages spatial politics and the politics of place. Undertaking Mission 3 of the project she addressed the means by which the politics of space and place are manifest and described at a local level. The heritage orientated approach of the Quarters project constructs an engagement with space as both sites to see and be seen within; for example landmarks of historical architecture at Queen's University or the social cultural hub of the Cathedral quarter. In contrast, O'Beirn engaged the city through processes of mental mapping, recollecting and spatial identification by using the Shuttle as a form of information booth, to harness and exchange vernacular knowledge of the city from its inhabitants and visitors.

Titled *Some Things About Belfast (Or So I'm Told)*, the work formed an extensive collection of place nicknames, urban myths, anecdotes and hand drawn maps, amassed through contributions made by anyone who entered the shuttle and shared their spatial knowledge, experience, and perception of Belfast. From the collation of names, streets and sites, long black paper columns printed in white text pair the common nickname to the specific locale it refers to; 'the chicken walk' meaning 'the easy route up Divis Mountain', 'the coolers' being the 'former outdoor pool now bmx track, falls park' or the 'dam back' as the 'alley at Glandore Grds., Ardoyne' (O'Beirn 2006: online). In a series of posters, line drawings and text reveal an intimate relationship with the city and contextualize its use. *Beakers Cave* refers to Black's Mountain, which flanks the city, used as a hide-out by children playing truant from school. *The Buried Virgin* refers to the urban legend of a statue of the Virgin Mary reputedly being buried, for a joke, into the foundations of the Rangers Supporters Club by construction workers; *Dogs on the Street*, depicting a stocky bull terrier, refers to the commonly used phrase where a particular knowledge is 'widely known at street level' (O'Beirn 2006: online).

The work curates a specific and distinct representation of Belfast where the relationships and politics of place can be read through a local vernacular level rather than the official formal and representative. In this way Belfast can be read as a place no longer burdened as merely synonymous with conflict, but a place enabled to be re-perceived, through the practices of the everyday, as a place of home. It highlights the elements that shape our identification with local belonging and exemplifies how the cultural character of cities circulates in the language used to define and describe them. Commonalities in the everyday use of the landscape, from playing truant to weekend walking in the Belfast Hills, come to the fore and while sectarian division and allegiances of conflict remain evident, they are inflected through irony. The site-specific method and the artist's role in engaging people over their particular relationship to the city were pivotal in activating a critical dialogue of exchange. O'Beirn comments: 'Visitors were served coffee and tea from Belfast story mugs and had the opportunity to peruse my collection and make their own amendments and additions' (O'Beirn 2008: online).

The impromptu one-to-one dialogue fostered by O'Beirn importantly echoed the manner and momentum in which vernacular knowledge is produced and circulates through the social fabric of everyday life. As the information accumulated, concurring and differing, even unfamiliar names for spaces emerged. In this way, the accounts of spatial experience prompted an intimate production of place and the relations of social space. For O'Beirn 'One of the unexpected outcomes of the project was that my personal knowledge of the city was tested by passers-by who gravitated to the Space Shuttle as a civic information point' (O'Beirn 2008: online). This configuration coupled with siting the shuttle in the heart of the city centre determined the diversity of visitor-participants that, as a result, realized the means of creating and encountering alternative perceptions and new insights of place. Moving mentally through well-known places and landmarks, each visitor mapped a local knowledge through an act of recollection, consciously remembering and navigating their contextual relationship to place, their recounted experience of the city, ordering memory into spatial form. In this way the mobile architecture of the Shuttle acted as the temporary material locus for memory where authentic narratives of identity and belonging were brought to the fore.

To encounter O'Beirn's Shuttle necessitated the act of walking within the city and, on arrival, engaged a process of mental re-mapping and re-tracing of routes, spaces and locales. The long-established relationship of movement and memory can be seen as deriving in Quintillion's architectural understanding of how memory functions; a model that stems from a 'narrative mobile architectural experience of site' (Bruno 2007: 21). First physically, then mentally re-traversing and re-visiting spaces that contained carefully placed images relating to what was to be remembered, recollection was spatially determined, the 'architectural promenade, enabled the process of image collection to generate recollection' (Bruno 2007: 21). From an anthropological perspective walking can further be seen to underpin the alignment of movement and memory. This is examined in the work of Tim Ingold, as he reflects on the practice of walking as remembering when 'journeying forward along a path or trail, one is also taken back to places imbued with the presence of ancestors.' He notes 'trails are routes to remembrance just as they are routes to knowledge' (Ingold 2000: 148). The peripatetic nature of remembering can still be seen in certain tribal culture; when elders and ancestors are to be recalled, the present generation walk in the paths and routes of their forebears; their longing is 'trailed' in a moving practice of remembrance. Mirroring the peripatetic, primordial processes of remembering, O'Brien's 'mission' realizes a critical re-presentation of place whereby intimate knowledge and nuance of experience permit new understandings of the relationship to place and the context social and cultural division. As a platform for the generation of spatial knowledge the Shuttle can be read as an effective and productive site of critical exchange and interpretation.

Territories of identity

A sense of place and the formation of identity are further consolidated through the repetition of performative habits in the spatial practices of everyday life. The concept of belonging is examined in the work of architect, curator and writer Neil Leach whereby he considers the means by which identification with place occurs through transposing Judith Butler's theory of performativity to a discourse of belonging (Leach 2005). Repetition suggests 'a way in which communities might colonize various territories through the literal performances – actions, ritualistic behavior – acted out within a given architectural stage, and through those performances achieve an attachment to place' (Leach 2005: 285). Understood in this way, meaning – like memory – does not pre-exist as though naturally embedded in space, rather it is through the habitual performances, rituals and cultural behavior in the spatial practices of our everyday movement that meaning and memory is produced. Leach further explains 'belonging to place can be understood as an aspect of territorialization and out of that belonging identity might be forged' (Leach 2005: 285). In this way processes of identification become spatially framed whereby the built environment serves as a platform for the performativity of identity producing attachment to place.

Definition of the respective identities within Northern Irish society are commonly understood to be Irish-Catholic-Nationalist and British-Protestant-Unionist, represented in binary opposition to one another.[4] Narrated and performed within the educational, religious, social and cultural establishment, collective identity is further reproduced by respective parades, celebrated rituals and commemorative events. In this way it can be argued the concept of otherness becomes mapped into the social, psychic and spatial fabric of place, which, under the conditions of conflict, inscribes the territorial boundaries of identity and belonging, allegiance and division. Conflict transforms the contextual use and functionality of space. Within the everyday environment, street corners, bus stops, playgrounds, shops or car parks become potential sites of traumatic event and target as much as municipal buildings, security forces or borders of defense. The domestic realm also becomes subject to attack; brutality or killing carried out in residences violate and destabilize the psychology of home as the space of ultimate privacy and security. Freedom of movement becomes curtailed and habitual spatial practices such as the daily routes to and from work or school become carefully chosen variable journeys that navigate territorialized boundaries of division and the interface of being in and out of place.

The transformation of post-conflict space most commonly begins with the gradual dispersal of the physical structures that managed the boundaries and defenses of division – security checkpoints, observation towers, fortified police stations – yet the clear demarcation of Protestant and Catholic areas remain highly visible. The physical parameters of peace walls and the representational boundaries of painted streets sustain the territories of community division and arguably reinforce fixed space cultural and political narratives of identity. Initiatives such as the whitewash of some sectarian murals render the appearance of non-segregated territory and where new-build private homes flank established housing estates, decorative perimeter walls and gates merely operate as aesthetically disguised boundaries of economic as well as sectarian difference. Evidenced through the spatialization of fear, the effect and reach of conflict is neither readily overcome nor erased.[5] Considering the functional importance of the daily environment, Brendan Murtagh highlighted the critical nature of spatial routes and movement to regeneration and development projects. Exemplified in the planning of city bus routes as fundamental to a transitional society he warned: 'the radial routes [. . .] pay little attention to the way in which segregation mediates movement and creates distinctive mental maps that marginalize engagement with the wider city' (Murtagh 2008: 23). The post-conflict landscape must navigate territories of identity, which at the most quotidian level underpins physical transformation and the possibilities of shared space.

In relation to memory, Leach draws upon Walter Benjamin's notion of architectural appropriation to reveal its critical agency. The accumulation of 'sensory impulses, traces of their reception, constitute a type of archive of memorized sensory experiences.' He suggests this establishes a 'background horizon of experience' [and as such] 'identification might be understood as an ontological condition consolidated through memory' (Leach 2006: 142). Memory orientated, the experience and memory of place becomes attached to the structures and sites of events as much as it is produced in the movements and performative habits of its use. When an 'archive of sensory experience' encompasses a prolonged period of conflict, movement mediates the effect of the past in the present and in alignment with the dialectic of remembering and forgetting, reciprocally locates memory within consolidated boundaries of place.

Memorialization

In parallel to regeneration, the approach towards memorialization frames the experience and countenance of landscape. The problematic of the conventional monument lies in its capacity to misleadingly convey the notion of historical closure. Memorialization should not cease the interpretation of complex political, social and historical issues wherein the monument can stand in false or appropriated representation of their resolve, thus negating the necessity of critically attending the past. Memory invested in the monument, rather than ourselves, subsumes the responsibility of remembering from the social to the material and in this way monuments potentially disengage us from history. The issues of memorialization return the question of what we want from memory and raise the contention of what is possible and expected from memorial space. Here, memory can be seen as subject of strategic 'claim'.

Examining the political instrumentality of cultural memory, Farell Corcoran points out that it is specifically 'sites of memory' that 'reflect non consensual interests in divided societies' where one can analyse how 'memory is politically administered [and the] struggle for the heart and soul of public opinion is played out' (Corcoran 2002: 63). The erasure or preservation of particularly contentious historical and political sites oscillates between the fear of creating sanctified shrines to martyrdom and the eradication from the representation of history. The seemingly irresolvable problem of what, when and how to remember reflects the limits of the material memorial and raises the challenge of enabling new ways of remembering and representing the past. By example, the commission to commemorate the 10th anniversary of the Omagh bombing marked a significant shift from the tradition of memorial architecture in Northern Ireland. Carried out in 1998 by the Real IRA, the bombing is defined as the most unprecedented and indiscriminate act of violence in the history of the Troubles.

Awarded to the artist Sean Hillen and landscape architect Desmond Fitzgerald, the collaborative project highlighted the importance of the sensory nature of spatial experience as it relates to the practices of remembering. The design brief stipulated the need for scale and the integration of the garden and bombsite. It stated the memorial was to achieve a sensory and spiritual experience, to promote contemplation, healing and perhaps most notably to represent Omagh as a progressive place, a town that had moved forward, able to build a future and identity beyond atrocity. Technologically complex, the design was realized through collaboration with astrophysicist Peter Gallagher and aimed to harness the properties of light, fundamental to the experience of contemplation and reflection. A single large rotating mirror continually tracks the sun projecting sunlight onto the column mounted small mirrors that represent the thirty-one victims. They in turn are configured to project the light upwards and re-connect in the adjacent street with the actual site of the bomb where a glass and crystal obelisk stands, continually illuminated by light. The design intends for a 'simple meditative space' and Hillen defines his idea as 'a gesture towards redeeming human values in the face of atrocity' (Hillen 2007: online). The installation sits within an oval contoured garden flanked by the River Struhle. The site spirals inwards to a central pool, the mirrored columns bordered by semi-circular rows of birch trees, and the shallow embankments planted with herbs and clusters of native flowers. The design recognizes the contemplative value of half-light. Shadows cut and play across the surfaces of the site, heightening awareness to the phenomena of light and its metaphysical properties. Invited to sit and pass time, the visitor can in turn observe the passage of time through the mirrored, moving trajectory of the sun.

The project was not completed without considerable difficulties. Its reception met with highly divided opinion, oscillating between a successful construction of contemplative space and remembrance, and unsightly obstruction and financial waste. Historically Omagh is a town of near proportionate balance between Protestant and Catholic communities. Peaceable and tolerant co-existence arguably became a factor in its selection as a target. Commenting on the effect of the bombing, Edna Longley, Professor Emerita as Queen's University, observed how it 'induced a universal recognition of common tragedy' whose first commemoration was 'an extraordinary statement about mass death' (Longley 2001: 236). Four years since its inauguration, the continued dissatisfaction and debates for amendments reveal the extent to which all memorials ultimately must bear a portion of failure. Inevitably configured under the condition of their time and the political pretexts that seek to anchor a consensus of memory, memorials are thus tasked with impossibility. Andreas Huyssen reminds us that the issue of their failing is not 'the imaginative ability or inability of artists, architects and designers, but rather the objective problems of a death zone in living memory' (Huyssen 2003: 159). The memorial in Omagh does not fail to meet its brief; rather, its contention reveals the paradox of what is desired of memory, and the means to evolve new and divergent ways of remembering. Traditional memorials stand to represent and honour those who have died rather than question why they died. Memorial architecture – as can be seen in the case of Liebskind's Jewish Museum in Berlin or Maya Lin's Vietnam Veterans Memorial in Washington D.C. – hosts the capacity to construct an affecting sensory and embodied experience of space that serves to generate recollection, contemplation and, moreover, inquiry. In this way sites of memory and memorial spaces can sit in parallel with reshaped city and townscapes, not commanded through the duty to remember but as discursive spaces that may critically attend the past and be permitted to explore the 'whys' of history.

Conclusion

Concepts of place premised on the emphasis of process and change, as opposed to a static setting of experience, propose that place is a neither fixed nor finished entity, rather it is always in a state of becoming. Derived through a phenomenological perspective this bears correlation within a postmodern global sense of place where it is frequently argued that the notion of fixed, essentialized space gives way to a dispersed sense of place whereby movement, flow and networks take precedence over attachment to place.[6]

The conditions of flux, speed and mobility make for contextual shifts of place and while it is evident that within the pluralism of globalized place shifts in identity occur, they do not necessarily ensue. An attachment to place, particularly when rooted in the subjectivity of specific ethnic, social, cultural and historical experience arguably retains identities. In the case of Northern Ireland, the primary challenge of the post-conflict landscape is the notion of shared space and a common text of place within community division. The complexity of resolve, while driven through the terms of the peace process and the negotiation of models of governance, is evidently anchored in representation and the politics of place and identity.

Evidenced in the work of PS² and the break from the traditional monument in Omagh the role of art and architecture is proved integral and fundamental to the processes of transition and post-conflict normalization. The critical practices of inquiry, reflection and problematization make for extended understanding and representation of the past in relation to the ongoing influence it extends in the present. Furthermore, it can be argued that visual and spatial practice can further problematize the instrumentality of memory to determine new practices and ways of remembering, rather than be reduced to merely serving an illustrative or therapeutic function. In the context of the post-conflict landscape, the present day is placed in a form of continued negotiation with the past, the events and affect of which gradually shift between the survivor and successor generations. In the complex process of spatial transformation, memory underpins the degree of construction, erasure and preservation of historical sites and landmarks, and while it may seemingly pose an insurmountable obstacle to progress, it may also serve as the connective medium between the place of the remembered past and the space of the imagined future. While the practices and very real needs of remembering are essential to the pro-generation of the future, we must be mindful of the instrumentality of memory, ensure not to enforce memory and remember to permit the time and space for forgetting.

Chapter 10

Helen Chadwick

The *Model Institution* and personal identity

Stephen Walker

Introduction: Helen Chadwick's changing address to architecture[1]

This chapter will explore the British artist Helen Chadwick's (1953–96) changing understanding of, and response to, the impact of the built environment upon personal identity. Chadwick's interest in architecture was enduring and multi-faceted; her work enjoyed a two-way relationship with architecture, although the phasing and modality of this relationship were complex. Partly motivating such work was her clear belief that while the impact an environment had on identity ought to be a two-way process, architecture's traditional approach, product and judgement system all too frequently imposed upon its occupants.[2] In its four sections, the chapter will examine four key areas of Chadwick's exploration through which she scrutinised and challenged this situation. These are: her early, built critique of the built environment; her conceptual expansion of the field of relationships between built environment and identity; the role of the body and the impact of scientific thinking on its deployment as metaphor that reinforces common sense acceptance of architecture and authority; and an unresolved call for ambivalence to be accepted within the conception and actualisation of our relationships with ourselves, with others, and with the spaces around us.

Chadwick quickly came to believe that architecture's imposition on its users had been able to continue more or less unchecked because of the extent to which architectural assumptions were commensurate and complicit with a broader world-view based empirically on Newtonian physics and conceptually on Cartesian philosophies. While she developed a way to critique this situation theoretically, she was concerned that the everyday was adequately explicable by such systems, to the extent that other approaches would fail to gain any real purchase there. This reflects the extent to which architecture appears to be common sense, something that we just take for granted and accept as the background to our lives, and our individual and group identities.

Her early projects made such imposition manifest. To introduce pieces that will be discussed in more detail later, *In the Kitchen* (1977) played on and played out a domestic scene conflating stereotypes of women's position in the kitchen with kitchen furniture and equipment; it was a full scale piss-take involving fabric cooker suits and other kitchen appliances, worn by female protagonists whose identity was here literally engulfed by these domestic appliances, and whose agency and capacity to act on their environment were clearly emasculated. Dealing with related issues in public rather than private space, *Train of Thought* (1978–79) and *Model Institution* (1981–84) were installations with associated performances and soundtracks respectively. These developed around Chadwick's close observations of the impact that an environment had on people, and articulated various issues that were to become central to her œuvre. What she explored in these pieces concerned as much the non-physical as the physical aspects of the interface between self and environment. While the physical environment might appear neutral, there are other institutional and individual factors that bear strongly on both individual experience and social interaction, and individual and collective identity.

In *Model Institution* most explicitly, she explored the environment's role in establishing and maintaining the individual's relationship with power: this piece was, in her words, a 'model of control + pressure' (HMI 2003.19/E/3:23). In her notebooks, Chadwick writes of the inter-connections across the interface between self and environment. There, and in the piece itself (and the piece was both a performance and a subsequent publication), she announced her interest in various overlapping readings of the same place that set this interface up as more than just a physical architectural boundary, exploring experience and identity beyond restrictive binary models that might be crudely aligned to the different sides of the glass screen at its centre (HMI 2003.19/E/2:26).

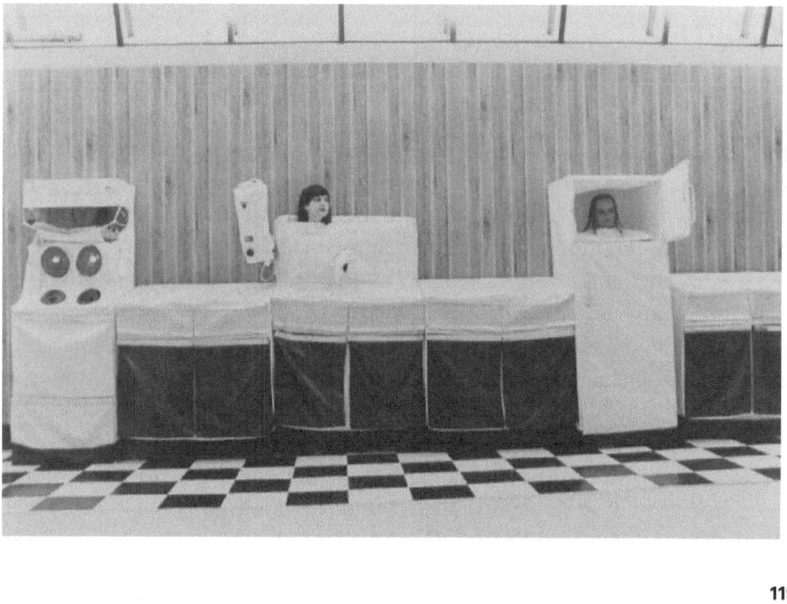
'e 10.1 Helen wick, *In the* en (1977).

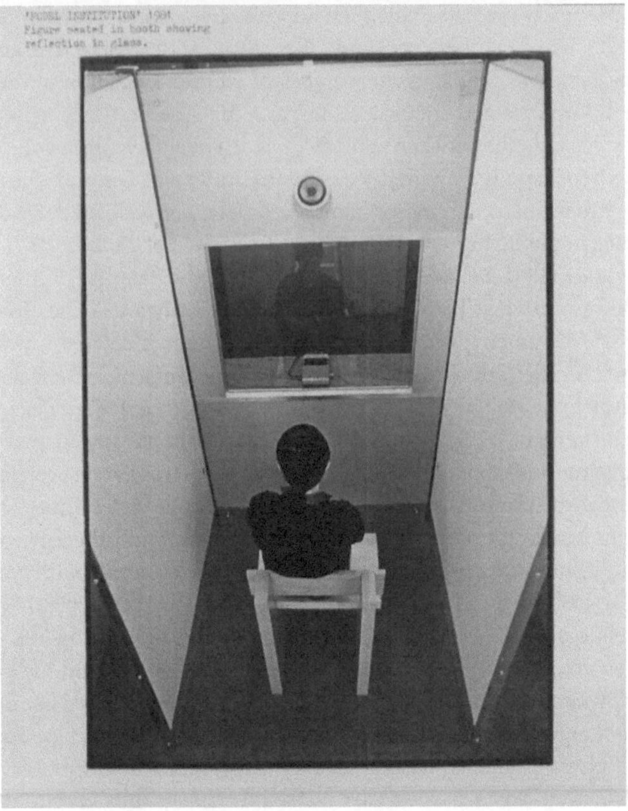

Figure 10.2 Helen Chadwick, *Model Institution* (1981–84) showing figure (Chadwick herself) in one of the five booths that made up the whole installation.

Now while she appeared quickly to abandon the outward, faithful reproduction of architectural environments after *Model Institution*, the issues she raised concerning the interface continued to occupy her, although they developed a good deal over the decade that followed. Architecture seemed more or less to disappear from her realised work at this point — marked by the major, operatic projects *Ego Geometria Sum* (1983–85) and *Of Mutability* (1986–87) — and it might be suggested that Chadwick's interest in and engagement with architecture tailed off fairly quickly as her career progressed. Her realised œuvre demonstrates a clear move away from architectural installation work towards more autonomous and poetic two- and three-dimensional objects. However, to imply she was no longer interested in the various political and disciplinary concerns regarding the relationship between identity formation and architecture is erroneous. Reflecting on her early work a decade later, she observed how she felt this marked 'a possibility for beginning to look at identity and the conditions where it collapses' (Chadwick 1990).[3] While it is beyond the scope of this chapter to pursue the trajectory of her interests in identity politics through her entire œuvre,[4] what can be undertaken here is a closer examination of the conditions of collapse as these were first rehearsed in projects such as *Model Institution*, for in so doing, their broader relevance for debates concerning architecture and identity can be suggested.

Buildings stronger than people

The notes she kept around the time of this apparent abandonment of architecture indicate her changing conception of both architecture as an institution (and this in both the sense of buildings standing for the broad institutional values they purportedly upheld or imposed, and which were played out so strongly in *Model Institution*, and also in the sense of architecture itself as professional and disciplinary institution) and of the role and potential of art to address such institutionalisation. In 1982, more or less coincident with this clear departure in her realised work, she observed the almost overwhelming impact buildings appeared to have on people:

> alienation v. loss of identity
>
> bricks and mortar remain, physically concrete yet memory + past life gone, evaporated
>
> buildings stronger than people, – more enduring, more real than lives led
>
> (HMI 2003.19/E/5:1)

While this can be read as a lament, and contextualised within her œuvre as the point where she gives up on architecture, I would suggest that it marks a moment when she stops banging her head against the wall of architecture by trying to tackle it head-on, and instead begins a more vigorous and complex investigation into how this apparent and problematic 'strength' could be addressed. While the detail of these investigations is too complex to cover here in detail, it is worth saying here that by the end of the decade, Chadwick had become far more upbeat about architecture, which suggests that her work in this area had made progress. Her considerations had moved from architecture as object to architecture as creative and experiential process where, in her own terms, the self becomes a building site around which these renovated architectural readings could take place: 'architecture grows corporeal and I am enfleshed' (Chadwick 1989: 109).

This broad reconsideration regarding how best to approach architecture is echoed around the same time by questions about the role and method of her own work. Just as she had begun to express dissatisfaction with her work's relationship to architecture, so she writes of her restlessness with artwork that operates within the status quo of the art institution. Her response to both these situations was to try to vary or broaden the parameters against which the work might operate, to try to get beyond established rules and relationships governing people's relationship with art or architecture. It could be said that both these aspects of her work move away from direct contestation towards more involved and creative engagement. Her feeling towards these institutions found voice in a passage in Arthur Koestler's book *The Sleepwalkers*, one of a number of related works she read around this time and which had a significant impact upon the development of her thinking. The changes she sought in her own work reflect her broad belief, following Koestler, that art (and architecture) were 'ripe for change.'

> The symptom that a particular branch of science or art is ripe for change is a feeling of frustration and malaise [. . .] This is the situation which provides genius with the opportunity for his creative plunge under the broken surface.
>
> (Koestler 1982 [1959]: 530)

Just as her lament about the strength of buildings gave way to the more celebratory assertion of *enfleshing* (Chadwick 1989: 109), so in a more or less contemporary remark alongside Koestler she saw an opportunity for a plunge under the broken surface of contemporary modern art.

Chadwick's interest in and close reading of Koestler's work can also help us to understand the shift both in her understanding of the relationship between architecture and identity formation, and in her artistic response. Her broad position regarding the architectural environment can be suggested by quoting Koestler's opening paragraphs of Chapter XIII 'The Glory of Man' from *The Ghost in the Machine*, a chapter that Chadwick annotated heavily in her own copy. Koestler argues:

> The activities of animal and man vary from machine-like automatisms to ingenious improvisations, according to the challenge they face. Other things being equal, a monotonous environment leads to the mechanisation of habits, to stereotyped routines which, repeated under the same unvarying conditions, follow the same rigid, unvarying course. The pedant who has become a slave of his habits thinks and acts like an automaton running on fixed tracks; his biological equivalent is the over-specialised animal—the koala bear clinging to his eucalyptus tree.
>
> On the other hand, a changing, variable environment presents challenges which can only be met by flexible behaviour, variable strategies, alertness for exploiting favourable opportunities.
>
> (Koestler 1967: 201)[5]

There are several important issues that can be pulled out of this passage, the most directly relevant of which is the relationship between the stimulus provided by an environment and the wellbeing or atrophy of those using it. While exploring this basic premise, there are other aspects that Chadwick's early projects draw attention to, aspects less easily identified with the environment itself but which should be taken into account because they influence this relationship, such as politics, discipline and control, and the balance between 'nature and nurture'. Finally, while she would align her own political and artistic project with Koestler's ambition to encourage flexible behaviour and individual agency, Chadwick remained troubled by the extent to which her own work operated through the recreation of monotonous environments.

From monotony to the triplet

Returning now in more detail to Chadwick's early projects such as *Model Institution* with Koestler's passage echoing in our ears, her installation (and the spaces to which it referred) can clearly be read as one of the monotonous environments that Koestler links to the mechanisation of habits and stereotyped routines. Contextualised perhaps as a response to the contemporary unemployment crisis and social disaffection of the early years of the Thatcher government, the closely observed characteristics that informed its stark architectural interface encapsulate the intimidation felt by the vulnerable within any institutionalised environment. But beyond the physical construction of the five booths that provide the setting for the audience's interaction with the project, Chadwick's entry in her notebook reminds us that she considered this 'An architectural sculpture for 5 voices' (HMI 2003.19/E/3: 23). This indicates an awareness of a greater range of complexity amongst the factors that contribute to such environments (and can move them from monotony to menace). As further notes made during the preparation of the piece indicate, Chadwick was mindful of less material ingredients:

> The Institution
> Light
> Heating – fanheaters
> Atmosphere
> Unit cubicles
> Surveillance cameras: phasing
> Control
> Work }
> Health: throat } Tapes
> Food } Files
> Beauty } Forms
> Fun }
> Scale distortions lengths heights
> Research government papers
> Optimum light + temp levels
> Surfaces
> muffled sound
> frequency of light-flickering tubes
> Cubicles: a sculpture for five
> bureaucratic menace
>
> (HMI 2003.19/E/3:2–3)

While the *architectural sculpture* faithfully repeated the kind of institutional architecture Chadwick had investigated, this far exceeded the simple reproduction of the architectural object or cubicle: the soundtrack was every bit as carefully researched and developed, and more nuanced than the architectural cubicles themselves. It played out both the audible aspect of the monotonous institutional environment – 'Tintinitus | Ringing in ears | Sound accompaniment' (HMI

2003.19/E/3:18) – while also relaying encounters with and resistance to the kind of mechanised, routine behaviour that Koestler suggests will result from monotonous environments, giving voice to those subjected to this kind of monotony. Preparatory notes sketched out the tone of these voices:

> 5 arguments with 5 different personalities/roles
> Adopt different degrees of aggression + sources of grievance [. . .]
> all of whom were
> . . . individuals trapped by various circumstances
> Angry Man vs. system/power
> Woman vs Man
> 5 conflicts. Burdened down by intolerable positions, crying out impotent unable to change anything, voices of passion
>
> (HMI 2003.19/E/3:31)

The soundtrack inclusion of the five voices staged a short moment of resistance and primed the audience to relate to the piece rather than leaving it uninhabited and somewhat abstract. Nevertheless, Chadwick's admission that these characters were 'burdened down [and] unable to change anything' (HMI 2003.19/E/3:27) acknowledged that the soundtrack, like the architectural environment of this piece, ultimately repeated the overall monotony that Koestler suggests will lead to the mechanisation of behaviour and the steady conversion of these people into automatisms. However nuanced and well observed installations like *Model Institution* were, Chadwick came to realise that institutional critique was for her a cul-de-sac, and that a direct attack on the monotonous environment was beyond the scope of her work.[6] Monotony, and the threat that this posed to individual identity and agency, could not be countered by more monotony. Monotonous, critical artworks introduced within the circuit of art galleries would not bring about any amelioration in the conditions of those represented in the soundtrack to *Model Institution*.

Towards the end of the notes she kept during the development of *Model Institution*, she included the first of what would become frequent reflections regarding both her own creative process and the broader relationship between art and life: 'remove art from its own institutions + legitimised manner of viewing/seeing – put it in unrelated institutions so functions under varying set [of] parameters' (HMI 2003.19/E/3:32).[7] Although for the most part she failed to follow her own advice, there are aspects of this imperative that are manifest in the works under discussion in terms of the ways in which she began to draw attention to the institutional parameters of viewing and seeing. Just as *Model Institution* addressed many non-physical factors that reinforced the geometric monotony of the architectural cubicles, Chadwick's earlier project *Train of Thought* similarly operated with an institutional architectural setting. In this case, it copied a bay of seating from a London Underground train carriage, over which several other factors affecting the experience of that environment were layered through a performance involving two characters, a female and a male passenger who were strangers to each other, and whose internal monologues were followed by the audience. While the actual performance anticipated the use of the soundtrack in *Model Institution*, it is interesting here to focus on the project's subsequent publication in AMBIT magazine (Chadwick 1979: 17–22). In addition to photographs of the performance itself, this article included annotated photographs of the architectural environment. The graphic style of these photos mimics orthographic architectural drawing (here, the cross-section) and also the presentation of anthropometric and ergonomic data in design guides used by architects and designers such as the *Metric Handbook* or Ernst Neufert's well-known *Architectural Data*.

In that article's Figure 5, entitled 'Anthropometry and Human Engineering' for example, the typical highly reductive presentation of 'average' human data is applied to a photograph of the two protagonists in the piece.[8] In addition to the conventional dimensional data such as 'Eye Height,' 'Elbow Height,' 'Buttock–Knee', and so on, Chadwick continues within this graphic style to replay some of the central themes of the performance. The internal monologue of the female passenger played out her rising fear caused by the presence of the man opposite, and was reinforced by the arrangement, character and isolation of the environment. As the male passenger's thoughts revealed, however, this fear was misplaced – he was not planning to assault the woman opposite, and

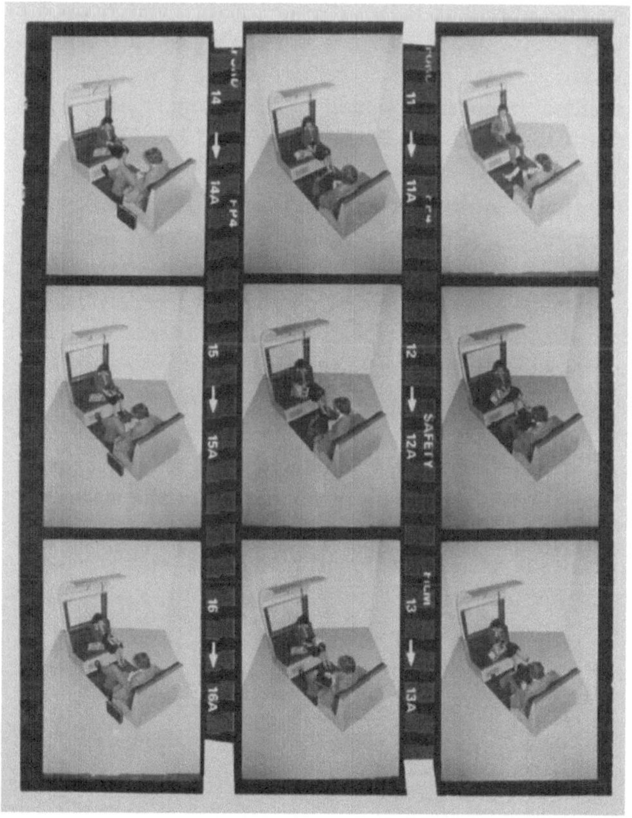

re 10.3 Helen dwick, *Train of ught* (1978–79). act sheet showing ral arrangement stallation, with protagonists in ecreated London rground carriage ng.

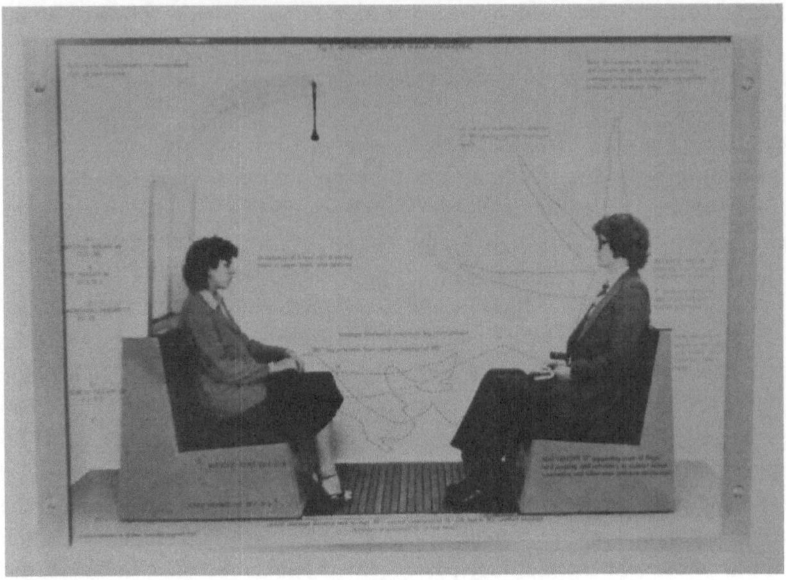

Figure 10.4 Helen Chadwick, *Train of Thought* (1978–79) published here in Chadwick, H. (1979) 'Objects' in AMBIT 81 ('A quarterly of poems, short stories drawings & criticism 17–22.

his thoughts were elsewhere, actually worrying about his work. This physical proximity between the two characters within the highly functional design of the tube carriage is quietly reinforced in the annotations to the AMBIT spread. 'Passage blocked if maximum leg room utilised', for example, indicates the potential zone of contact between the two overlapping, dotted anthropometric figures and echoes the tension played out in the performance of *Train of Thought*.

This potential physical overlap and interaction is one of several layers that Chadwick believed affected the experience of space and personal identity and that were mediated across an interface between individuals and environment. While this was clearly important for *Train of Thought* as a project, it remained a consistent motif through much of Chadwick's œuvre. Her early label for the dynamics that occurred across this interface was the *Triplet*. It is worth exploring the *Triplet* a little further here, as this can help to articulate several of the key questions that Chadwick consistently pursued, and various ways in which this ongoing exploration developed architectural issues that side-step the problems she acknowledged regarding the strength of buildings and the limitations of straightforward institutional critique artworks.

The *Triplet* is mentioned in a notebook she kept between March 1975 and 1978, where she refers to the preparation of the images that appear in the issue of AMBIT just mentioned.

Series same image
Furniture area
Arch. drawing ergonomics of figure
Narrative
"Spaces of sensations

Spaces of desire
Spaces of negation"
Triplet object / action | architectural area plan | diagram paths to destiny
Façade / cross section / (sex) action
Annotated markings in real space
Manifesto: plot / fantasy desiring a space to exist.
HERE→space desiring a plot

(HMI 2003.19 / E / 2:26)

The ambiguity or tension across the *Triplet* between the architectural arrangement of objects in space and the potential animation of that space by people marks not only a temporal dimension acknowledging possible changes over time (discontinuous to be sure, set out between the prior architectural plan of the area, the actual action that takes place there at any point in time, and a notional future point or potential projected by the users and marked by Chadwick as destiny or fantasy) but also the superimposition of aspects of the complex power relations in play prior to and between them. The 'annotated markings in real space' (HMI 2003.19 / E / 2:26) become legible on the images in AMBIT, where they refer to the unrealised, unwelcome potential for interaction, harassment and conflict – 'Passage blocked if maximum leg room utilised' (Chadwick 1979: Fig 5) – but they also mark Chadwick's more general criticisms of the impact that rationalised approaches to planning can have on the actual experience of a space, and indeed her enduring preoccupation with the relationship between science and our understanding of the world and our place within it.

The juxtaposition of anthropometric data with the realised spaces of the tube carriage seating goes some way to qualifying her lament that 'buildings stronger than people' by drawing attention to the rules and guides that are prior to, and often more enduring than, the bricks and mortar (or in this case the carriagework and upholstery) of our everyday environments, and which have a significant impact on how people can behave there. The anthropometric overlay gestures to various techniques used by those in power to maintain their hold on that power, to police cultural norms and behaviour, to impose identities on those subject to these spaces. Reintroducing terms introduced earlier, the institutional parameters of viewing and seeing are here writ large rather than remaining invisible: for Chadwick, the revelation of such techniques marks the point at which agency could be exercised, and power circumvented or challenged.

While this applies at the scale of the seating booth and the interaction between passengers, it also operates at the scale of the individual human body. Here again the illustrations from AMBIT play on the differences between the individual actors in the piece and the reductive anthropometric figure, which reduce the complexity of human actors in extended networks of relationships, each with their own personal histories and futures, to an object out of time. At this individual scale, identity is frequently tied to a bodily motif or conceit, a scientifically determined, statistical and geometric representation which provided Chadwick with another institutional parameter of viewing and seeing to be challenged.

Common sense and the challenge to the bodily metaphor in architecture

The significance for Chadwick of geometry's relationship to the body and space in particular, and of the relationship between science and our experience of the world more generally, cannot be overstated. Her research and theoretical work explored the ambiguity and potential conflict between a rationalised 'average' person portrayed in the anthropometric figure and the actual identity of the individuals who used the space (and played out in greater complexity in the soundtrack of *Model Institution*, for example). Although the early projects discussed here clearly have as their target the impact of excessively deterministic approaches to producing architectural space, and the concomitant reduction of individuals to a lowest common denominator anthropometric diagram or bureaucratic statistic, the various dimensions of the interface brought together in the *Triplet* sustained her considerations of what was to become one of her most enduring concerns: how to reconcile or negotiate between the Newtonian-Cartesian world-view and those offered by more recent work in psychoanalysis, identity politics and modern physics.

While her work pursued the consequences of this tension in many areas, Chadwick believed this tension was most easily ignored in the production and experience of the designed environment. She frequently observed the ease with which traditional Newtonian-Cartesian approaches can account for everyday experience of the environment. While acknowledging the importance of such a common-sense understanding, her work demonstrated complex and sophisticated considerations of how the insights of modern science might bear on our relationship with our surroundings. Issues that developed around her reconfigured understanding of identity politics suggested ways to counter the Newtonian strength of buildings by providing a clearer understanding of how and where architecture operates with respect to its traditional role in policing identity formation, and offered an alternate knowledge to inform transformative strategies that promote individual identity and agency.

The implications of Chadwick's œuvre call the traditionally clear distinction between the design of space and the experience of space into question. Indeed as Chadwick's considerations imply, there is a central reciprocity that links not just the experience but also the broader role of the user with the creative process. Notes made around the time of *Train of Thought* indicate that she had begun to research links between body and space that would operate beyond the traditional architectural measure of and metaphors linking body and building (and sometimes city), which had frequently reduced both to a proportioning system, sometimes to a metaphor for the disposition and hierarchy of spaces and uses, or to mechanistic metaphor for circulation.

Her early project *In the Kitchen* suggests that while she acknowledged these various metaphoric links, her initial approach was to address them indirectly through the critique of gendered power relations that the project targeted. *In the Kitchen* anticipated some of the central issues of both *Train of Thought* and *Model Institution* in a domestic scale and setting, playing on these links by conflating body and domestic machine (as Figure 10.1 shows, she and others dressed up as an electric cooker and other domestic appliance garments such as a washing machine, a fridge, and other kitchen storage units) in order to comment on the power relationships that she observed in domestic space, in who designed and made these machines and who actually used them.

Her research around this time had begun to investigate more nuanced uses of space that would be linked together in the *Triplet*. This fuelled her growing awareness of and interest in the politics of viewing, overt and covert uses of space, and the complex roles played by clothing in allowing simultaneous (social) conformity and contestation, all of which are manifest in her architectural installation pieces.[9] In addition to these realised works, ideas for other unrealised installation pieces played out the metaphoric application of such issues more literally; her concurrent interest in the militant feminism of Valerie Solanas' *S.C.U.M. manifesto*, for example, is reflected in this proposal for an installation project:

> *Womb Room* / environment / Gallery / brothel
> Rounded corner: non angular . . .
> Internal surfaces: feathers, eiderdown / quilting / upholster / plastic shower curtaining wallpaper
> Door entrances: fanny – lips, clit mound venus: hair
> Walk into vagina→womb
>
> (HMI 2003.19 / E / 2:1)

In this proposal, as in many subsequent instances, Chadwick closes the gap between the two ingredients to such an extent that the very operation of metaphor collapses, and thus the credibility of the longstanding architectural metaphor that links body and building is undermined as Chadwick short-circuits the operative distance required between 'body' and 'building.' By proposing to build a womb-room rather than a womb-like room, she draws attention to the risible assumptions that underlie many references to the latter, and to the broader objectification of the body assumed by such metaphors generally. She undermined these by proposing to literally build them as objects.

Although this unrealised womb-room proposal was very much a one-liner, Chadwick's appreciation of metaphor endured and developed (though she never entirely tired of the shock-value of this conflation and reversal). Writing around 1981, she continued to consider the metaphoric role of the gallery, though develops the potential of this role by expanding the range of issues the metaphor is charged with linking.

> The gallery becomes the memory
> "   is the brain
> Walk in space = getting inside artist's head
> Curiosity is motivation
> In gallery look at architectural drawings plotted from photos of some of more relevant buildings [. . .]

> Contrast of formal drawings of buildings – idealised – with actual photos
>
> (HMI 2003.19/E/5:9)

The importance of the shift between the gallery-as-womb-room and this later approach to gallery-as-brain lies not in the move from body to mind (or with any suggestion of more worthy or higher aspirations this might imply) but rather in the attention Chadwick pays to expanding the modes of engagement between artist, work and audience. Here, the gallery space is charged with mediating between artistic intention and observer's experience, pursuing metaphoric links between the gallery and the head, the brain, and memory. Echoing concerns played out in *Train of Thought*, and expressed theoretically in the *Triplet*, this expanded engagement works through the build up of, and interaction between, a number of layers which call into question the authority assumed by the idealised realm by clearly juxtaposing the ideal with reality, which here is legible in the contrast between formal drawings and actual photos.

The development of her theoretical position reflected her growing conviction that our worldview needed to account for a broader understanding than that peddled by science (or at least the reductive quasi-scientific design of anthropometric determinism), one that could embrace more illogical and dysfunctional aspects, in order to contest the rise and authority of a narrowly empirical scientific position that she believed had come to dominate many aspects of life where it should hold no sway. Faced with a paucity of holistic thinking that might support her interest in combining both ideal and real, Chadwick followed Koestler in drawing a distinction between the older terminology 'philosophy of nature' and the newer 'physical science'.[10] Reconsidering the former, she believed, offered ways to expand the contemporary environmental debate beyond the limitations of empirical science. In perhaps the clearest statement of her position, she emphasised the importance of redressing this balance, and what was at stake:

> Science not only way of gaining access to truth of universe. Access nature of reality thro' non scientific means.
>
> What left out – to heal divide of art science– might be consciousness/ subjectivity/ lived time IDENTITY
>
> (HMI 2003.19/E/8:143)

In conclusion: the value of ambivalence

A decade after identifying the opportunity for a creative plunge under the broken surface of contemporary modern art, she talked of her work as involving 'a leap across.' Writing in 1992, Chadwick demanded that we should:

> Take [my] works as clues to the value of ambivalence, of doubling, of a leap across, of the helix of pleasure around language. Pleasure is

> the field, the reward when the singularity of want is transformed into simultaneities that cannot be divided or closed.
>
> (Chadwick 1992a: 73)

Just as her later realised works combined monstrous and beautiful, taboo and accepted, entrails and flowers, shit and chocolate, her written advocacy of pleasure must not be taken simply to replace the gritty social realism of her early, more militant politics. Indeed, the value she placed on ambivalence, simultaneity and openness had by this time become central to her position regarding identity politics, art and architecture.

In contrast to the stand-off between occupant and environment critiqued yet recreated in *Train of Thought* and *Model Institution*, the increasing importance she placed on ambivalence (both theoretical and in her later realised work) put the potential of the *Triplet* into play in ways that attempted to exceed the certainties of the Newtonian world, and the Cartesian separation between mind and body that acts as its corollary. Chadwick sought to promote the identification and consideration of this interface-as-triplet beyond that of the clear boundary associated with traditional distinctions between body and world, mind and matter, and so on. All her later work was to recognise, to champion, the value and ubiquity of indeterminacy and the impact this realisation could have on identity formation and maintenance, in contrast to that permitted within received epistemologies and architectures.

Her later, more poetic works attempted an alternate address to the *Triplet*, to provide an environment or experience in which overlapping ingredients could open onto less monotonous environments. While this can be read as a fairly clichéd ambition and a situation that the architecture profession, or the construction industry more broadly, have failed to realise, it was accompanied by a more thoroughgoing change in how she, or we, could consider the setting and ingredients of identity formation and maintenance. The most important aspects of this were twofold: firstly, her insistence that in addition to the physicality of our built surroundings, the invisible, institutional parameters have an equal strength and are perhaps harder to identify and challenge. Secondly, that this combination is itself subject to continual flux and challenge, and that while this frequently remains (or is maintained) out of sight, it can be taken up and appropriated by those usually subjected to the interests of others. As a result, the comfortable categories around which identities have traditionally been discussed and considered – namely the discrete entities of the social and the individual, the building and the body – can be consumed within this broader field.

Acknowledgements

I would like to acknowledge the support of the Henry Moore Institute in Leeds, and in particular to thank the archivists Victoria Worsley, Ian Kaye and Claire Sawyer. I would also like to thank the editors of AMBIT magazine for permission to reproduce Chadwick's article.

Part Three

Representation and identity

Chapter 11

Identity, culture, and urbanism

Remarks from colonialism to globalization[1]

Nezar AlSayyad

Colonialism, nationalism, and globalization

The twenty-first century consummated the revolutionary developments that have been occurring in the world since the 1970s. Trends like the transnationalization of capital, the internationalization of labor, the steady increase in global trading and communication, and the ensuing competition between cities have led individuals, businesses, industries, and governments to attempt to position themselves globally (King 1991). It follows that in a globally compressed world, constituted of national societies that are becoming increasingly aware of their ethnic and racial roots, the conditions for the identification of individual and collective selves become very complex (Robertson 1991). Any theory of *globalization*[2] must take into account a necessary recognition of the distinctive cultural and unequal conditions under which the notion of the 'global' was constructed. Globalization becomes difficult to comprehend without recognizing the historical specificity of traditional cultures, their colonization, and their later emergence as nation-states.

At the heart of all of these issues is the question of *identity*. Of course, the connection between the identity of a people and the form and culture of the cities they produce has been the subject of much research. Family, ethnicity, religion, language, and history have all been identified as identity-constituting elements that are handed down in a process normally referred to as 'tradition' (AlSayyad and Bourdier 1989). Tradition is based on valuing constraint. In a technological world of limitless choices, the conflict between the traditional and the modern in urbanism becomes unavoidable. To cast the traditional as good and the modern as bad, or vice versa, would be naïve. As much research has demonstrated, the traditional/modern dialectic is very problematic. Dichotomies such as East vs. West, North vs. South, First World vs. Third World, core vs.

periphery, and developed vs. underdeveloped nations may be considered artificial categories. An approach that takes into account the discursive constitution of *cultures* would recognize that societies are constructed in relation to one another, and that they are produced, represented, and perceived through the ideologies and narratives of situated discourse (Wolff 1991). For example, Stuart Hall (1991) has observed that the dominant term in each of the above dualities (West, North, core, etc.) is mainly defined in difference, constructed in opposition to the other; none constitutes a monolithic, preexisting real subject itself. The subordinate term (East, South, periphery, etc.) is equally an invention, produced in a variety of postcolonial and anticolonial discourse. The only purpose of these dualities for scholars is to force an articulation of theoretical positions beyond the realm of binary opposition, in the process opening for discussion the complex dimensions of what is being categorized – in this case the underlying political bases of urbanism.

In studying the relationship between these two realms and its effect on the identity of people and their cities, we can discern three historic phases: the colonial period, the era of independence and nation-state building, and the most recent phase of globalization. These phases have been accompanied by three respective urban forms: the hybrid, the modern and pseudo-modern, and the postmodern.

Before the era of colonialism in most of the developing world, settlements largely took the form of traditional communities in preindustrial and often insular conditions. Although some forms of economic exchange occurred between this world and that of the developed world, curiosity about the 'other' was limited. Indeed vernacular forms of settlement were shaped more by sociocultural concerns and the surrounding natural environments. They also reflected, possibly at the subconscious level, the identity of their inhabitants.

Colonialism and the hybridization of settlements

Around the middle of the nineteenth century the world witnessed the rise of modern industrial capitalism and the emergence of organized political dominance represented by colonialism. Under this colonial paradigm, the world became divided into two kinds of people and two types of societies: powerful, administratively advanced, racially Caucasoid, nominally Christian, and principally European dominant nations; and powerless, organizationally backward, traditionally rooted, and mainly nonwhite dominated societies (King 1991).

The paradigm shift from the traditional to the colonial created a relationship of unequal cultural and socioeconomic exchange. Moreover, if one analyzes the issues of identity in the cities of the developing world, one must consider this fact and understand the processes by which identity was violated, ignored, distorted, or stereotyped throughout history. Artistic representations of colonial enterprises often depicted the colonized subjects as distinctly 'other', often in positions that were physically subjugated in relation to the figures of the European colonizers.

Here the problem of representation becomes central, because such artistic creations contributed to the creation of the 'other' and enlarged the gulf between the dominant and the dominated. Once the 'backwardness' of the traditional population was established (at least in the minds of the great mass of citizens in the colonials' motherland), reform was legitimated. Colonial regimes proceeded to eradicate the ethnicity of the colonized cultures. This series of events necessarily had an effect on the physical fabric of cities; everywhere they went colonists introduced their own brand of settlement.

The colonization process affected the overall planning model that determined the pattern of urban developments in both the core and the peripheries. This was the era when modernist ideas followed from countries of the North to the South. Ironically, in the 1950s and 1960s when the countries of the South launched their wars of liberation and independence, the North resorted to an age-old urban strategy. Thousands of traditional villages were destroyed in order to regroup the population in checkerboard resettlement towns under the banner of modernization. This uprooting operation was meant to break the subversive influence of the rebels. There was little desire to improve conditions for the local population.

The colonial era thus resulted in a hybrid urban condition. A certain architecture and urban language began to emerge that, at least at a visual level, unified the lands of colonial empires. For example, variations of the bungalow, a hybrid type of dwelling introduced by the British in India, began to appear all over the British Empire, making it difficult to identify its origins.

Nationalism, independence, and modernity

When the people of the dominated societies started to rebel against this colonial world order, they had little conceptual language to employ in their drive to establish sovereignty. Usually they were forced to use the terms of the existing order, with all its baggage of physical realities and ideological constructs like the nation state. Groups of people living in one region under a colonial power (but of different religions, languages, ethnicities, and traditions), sharing little more than a colonial history, had to band together to achieve this new, 'more advanced' stage of independence. The new political bodies highlighted what few commonalities existed, whereas they suppressed differences in pursuits of the noble goals of freedom. A national identity based on shorter-term political interest and the ideology of struggle emerged as the driving force behind most nationalist movements. Once independence was achieved, the glue that bound together the various groups no longer held. Indeed, the events of the late 1980s and early 1990s in places such as the former Soviet Union and Yugoslavia are a testament to the true associations of these people, as ethnic origin, race, and religion are reemerging as the prime definers of collective identity (AlSayyad 1992).

This second phase of independence and nationalism did not necessarily improve the quality of cities in the developing world, nor did it resolve the conflicts that plagued the traditional settlements of those societies. During the era of colonialism, important and irreversible decisions were made that affected the production of the built environment. In the Arab Middle East, for example, new building codes requiring setbacks (based on Western norms) forced the traditional courtyard house out of existence. New construction often took the form of banal single-family dwellings that were unsuited to the climate. Because of a host of cultural complexities, however, traditional people often prefer the modern units, despite the need for bizarre adaptations (such as building into the street to protect possessions). In a society that cherishes privacy, some people had to build freestanding walls as high as forty feet to shield themselves from their neighbors. In some countries, entire efficient systems of construction were abandoned because they did not suit the modern era. The urban system fell grossly out of balance, and the urban environment of many developing societies became pseudomodernized.

The puzzling image of a lamppost popping out of an apartment building in Al-Khobar, Saudi Arabia, is a good example of this imbalance. The lamppost appears to be submerged in the building, its neck emerging from an upper story to shine down upon the street below. Sometimes the base of the lamppost is visible at street level, only for its upper reaches to be consumed by the neighboring apartment block. At first glance, one might wonder if this is not some new type of postmodern architecture. In reality, as Akbar explains in *Crisis of the Built Environment*, such infringements on public property are common, because in traditional cultures people often are used to a more fluid relationship with the public realm, taking over parts of it without necessarily violating it (Akbar 1988). Modern codes and the presence of utility lines deny this traditional freedom. Perhaps due to both disrespect for the law on one hand and fear of the authorities of the other, building owners often resolved the conflict by incorporating the utility pole into their private space without interfering with its public function. Here there are many other examples of similar arrangements that have emerged from either ignorance of the regulations or failure of coordination among regulatory bodies.

The obsession with modernity that accompanied the early years of nationalism and independence has preoccupied most governments in the developing world. As a result, the Western pattern of urban development has continued to serve as the reference for indigenous populations, particularly the urban middle class who stepped in after independence to run the bureaucracies of developing countries.

During this time, the construction of public housing was seen in many parts of the Western world as an approach to achieve social justice. Among these efforts was the now infamous Pruitt-Igoe housing project in St. Louis, Missouri. Built as part of the urban renewal movement in the 1950s, the project was meant to house low-income residents of the downtown area. Yet after a few years the project deteriorated dramatically and attempts to improve living condition and lower the crime rate in its residential towers failed despite the allocation of millions of dollars in rehabilitation funds. By the early 1970s, the effort was written off as hopeless and the project was demolished by the government in an attempt to eliminate the social problems it represented. The demise of Pruitt-Igoe was considered by some to signal the end of the Modern movement in architecture (Bristol 1991).

Despite such public failures, the international influence of modernism was strong, and Western public housing models were often copied in the developing world without question. Developing world governments, however, used public housing projects as an instrument of nation-building in an attempt to buy the allegiances of the citizenry. Thus, at the same time as Pruitt-Igoe was being torn down, projects were being completed on the outskirts of several cities in the developing world with mixed results.

In some places, this public housing was reappropriated by the people. Egypt serves as a good example; under the socialist and nationalist regime of Gamal Abdel Nasser thousands of public housing blocks were built across the country. Suffering under the usual problems of such development – lack of maintenance, empty and unused spaces, the need to accommodate expanding families – many residents took matters into their own hands. Thus, one resident would decide to add a room to an apartment on the ground floor. Once built, this would provide the resident above with an excellent balcony. What then would prevent the upstairs resident from making it into a room as well? Of course, the usual response on the part of the government would be to send a bulldozer to remove such infringements on public property. However, residents also remained one step ahead. They often erected a mosque in the path that a bulldozer would travel; knowing that government would not hesitate to demolish residential infringements, but would never dare demolish a mosque. Thus, as informal additions continued, the original blocks would be absorbed. This example further shows the tendency of citizens of developing countries to resort to the power of religion to achieve their goals. Together with ethnicity and race, religion is becoming one of the preeminent forms of community identification in both the developed and developing worlds.

Globalization and the dismantling of the national project

Globalization is the third phase in the relationship between the dominant and the dominated. Life in formerly colonized societies arrived at an era in which the search for and reconstruction of identity had become paramount. Now that independence has been achieved and the dust from the struggle has settled, problems of national and community harmony have begun to surface. Where these issues have been resolved, religious and political fundamentalisms have begun to flourish. To understand the impact of these forces on urbanism, closer attention must be paid to the difficulties associated with defining national identity. The primary elements of nation identity – race, languages, religion, history, territory, and tradition – have always been essential but unequal components in its formation. The political units formed as nations after World War II were expected to be homogenous entitled with common culture. However, the reality was otherwise, because these nation states were mainly put together by international deals that displayed little interest in the will of the people who inhabited these lands.

Of course, national identity as perceived by a government is inherently tied to the image a movement wishes to project in the national arena. In the sociologist Wallerstein's view, the state, through its monopoly of policies and recourse, can create a national culture over time, even when it lacks one to begin with (Wallerstein 1991). However, can one design national identity? Moreover, can this national identity be designed by a foreign power? These are questions often faced by the politicians who governed the newly independent states. Architecturally speaking they were also faced by the architects and planners who worked for those politicians. Since cities are composed of forms that convey specific meaning, it is important to examine this physical dimension.

The relation between architectural form and national identity becomes especially important in terms of new 'monumental' architecture. Many new image-making architectural projects in the global south rely on star architects to somehow interpret a country's national identity through the built form. Foreign architects like Jørn Utzon and Louis Kahn were often charged with this particular task, and their resulting structures became symbols for the nation, even when their forms were more appropriate for Western cultures. About this ambiguity, Lawrence Vale, a historian of nationalism and the built environment, has written, 'At what point, one may ask, did the pyramid become an Egyptian form? Like the pyramids of Giza or the Eiffel Tower, the Citadel of Assembly may someday be seen as being quintessentially of its country as well' (1992: 127).

Nevertheless, there is much more here, as Benjamin Barber (1995) points out in his appropriately titled book, *Jihad vs. McWorld.* Many nations in this new global order want it all, despite the obvious contradictions. Such nations want the veil or the rule of religion, but they also want the World Wide Web and Pepsi Cola. Tim Mitchell (2002), on the other hand, argued that Jihad is not antithetical to the development of McWorld, and that McWorld is really McJihad, a necessary combination of a variety of social logics and forces. Thus, Third World countries evolve their own local appropriations of many First World practices. Hence, instead of Coca Cola, they now have their Mecca Cola named after the holy city. Meanwhile, for its part, the First World in this global era seems interested in consuming the cultures and environments of Third World societies. First World nations are often the main advocates for and financial patrons of the preservation of Third World built environments as part of what they define as 'universal' heritage – even when those countries 'natives' (particularly in the Islamic World) do not recognize its historic value.

A wealthy bloc, which often feels a sense of guilt and responsibility toward its former colonies, the First World has also tried at times to maintain or assist in preserving the dying or disappearing lifestyles and traditions of undeveloped peoples and places. Yet it is also often the case that First World organizations, foundations, and governments have engaged in such efforts while at the same time condemning or rejecting much of the social and political practices of the societies whose traditions they claim to want to preserve—especially when it diverges from Western standards of human rights and gender equality, as is the case with Muslim countries (maintaining women in traditional dress performing traditional functions to sustain the desire of tourists to see them is but one example of this paradox). It is still all about identity, culture, and form.

Of course, identity cannot be based on some myth from pre-colonial times. Many nations and their planners have resorted to a past around which identity may coalesce as a solace against the perceived dominator. Architectural historian Gwendolyn Wright rightfully warns, 'The past cannot simply mean a retreat to a golden age before the European, before modern industrialization, for these factors have changed us irrevocable' (1991: 311). Respect for the past in the developing world must include accepting and coming to terms with the colonial urban legacy.

If one accepts that national identity is a social construct tied to temporal events, as Louis Snyder (1954), the theorist of nationalism, has repeatedly argued, then it follows that the urbanism that accompanies it can only symbolize national identity as observed by a single individual or groups of individuals at a specific point in time. One must also recognize that there comes a point in life of all formally colonized peoples when they must cease to perceive their history as colonial and start absorbing this heritage as their own. When is this point reached? When did the forms of the British colonizers, for example the colonial house, become the vernacular heritage of the eastern United States? When did the Spanish colonial settlement forms in many parts of Latin America emerge as the traditional architecture heralded to tourists by these societies today?

The problem of national identity is further complicated by extensive global economic exchange. Not only do nations have to mediate between pre-colonial and colonial legacies, between the traditional and the modern, but they must also deal with the effects of globalization and the New World Order. *Globalization* here refers to the process by which the world is becoming a single economic entity, characterized by information exchange, interconnected modes of production and exchange, and traditional flows of labor and capital within a predominately capitalist world system. It would be convenient to adopt modernist Giddens' view that globalization has introduced new forms of world interdependences, 'in which once again there were no "other"' (Giddens 1990: 37). However, since capitalism thrives on the construction of differences, such an economic universalism, under the confines of a world constituted of national units, can only lead to further cultural division. Culture, then, becomes the globally authoritative paradigm for explaining difference as means of locating 'the other' (Robertson 1991).

Indeed, the considerable migration from the former colonies to the lands of the former colonizers and the infiltration of ethnic subcultures into mainstream First World Western societies cannot be dismissed. In fact, this phenomenon has often been the cause of social conflict, as these local subcultures have often resorted to ethnic, racial, or religious allegiances to keep from being swallowed up by the majority culture. The current struggle of multiculturalism and gender politics in the United States may be a good example of an attempt to embrace difference as a fundamental constituent of national identity. Ironically, while the national identity of the former colonizers is undergoing major change, often becoming more inclusive, the national identity of the formerly colonized nation is moving in the opposition direction, often becoming more exclusive and more directly linked to national origin or religious association. Indeed, the twentieth century has witnessed the return of states where belonging to a particular religion or ethnic identity is a prerequisite for nationality (AlSayyad 1992).

National identity is always a process of transformation and flux. While the often contradictory forces of globalization are playing havoc with traditional loyalties and values and challenging the older ideologies and practices, a single 'world culture' remains a distant prospect. Two controversial works of art that illustrate this position and demonstrate the complexity of the postmodern era are Salman Rushdie's (1988) novel *The Satanic Verses* and Saddam Hussein's Victory Arch.

Rushdie, a nationalized British citizen of Muslim-Indian origin, wrote his novel in 1988 employing analogies and using names drawn from the development of Islam. Although hailed in the West for artistic merit, the novel offended many Muslims in India, Pakistan, and Iran, resulting in the infamous Iranian death sentence against Rushdie. As historian Janet Abu-Lughod points out, events such as the Rushdie affair bring home 'how globalized and yet unglobalized culture has become' (Abu-Lughod 1991: 136). Had Rushdie written *The Satanic Verses* before the compression of time and space that now characterizes global culture exchange, it might have caused few ripples.

In similar fashion, symbols that new states inherit from history are often trapped in their old garb and an attempt to create new ones only produces images without context. In this light, Saddam Hussein's Victory Arch, shown many times on television during the Gulf War, acquired many negative connotations in the West. Ironically, these connotations were derived totally out of context. Consisting of two forearms and fists representing the arms of the leader Saddam, the arch was built in 1989 to commemorate Iraqi victories in the Iran–Iraq war.

Samir al-Khalil, the Iraqi dissident and the author of *The Monument*, has compared these enlarged arms to gigantic 'bronze tree trunks . . . [that] rise with their firmly grasped swords to an apex forty meters above ground. War debris in the shape of five thousand Iranian helmets taken fresh from the battlefield are gathered up in two nets' (1991: 18), and are embedded in the concrete base which supports the arms. Al-Khalil considers this monument a reflection of a bloody dictatorial regime. In truth, it is difficult to defend such an abominable piece of art or architecture, but, understood in its proper context, the arch is at best a product of popular culture, and at worst a monumental piece of design kitsch. As the gate for a military parade ground, might such an arch not have seemed almost appropriative? In addition, in terms of design, the arch builds on the role of the sword as symbol for honor, valor, and pride in Arab culture. A similar form of two interlocking swords is, in fact, the official seal of Saudi Arabia, a country that participated in the alliance against Hussein in the first Gulf War. The fact is that Western judgments, though justifiable in political terms, have been uninformed because of content.

Such cultural conflicts show that meanings can be easily lost in cyberspace and that careful dissection is necessary in order to understand the effects of a single-world system on cities. Furthermore, in every person it is likely that two conflicting sentiments exist toward culture. The first is to resort to culture and tradition out of fear of change – change that in and of itself may be inevitable. Protectionism against the unknown does, however, turn into fundamentalism. The second sentiment, characterized by interest in the culture of the mysterious 'other', emerges from a very different feeling: the desire to have the choice to merge with the 'other' and share in a wiser or a different collective consciousness. The tremendous movement of citizens across borders and the rise of protected ethnic minorities demonstrate that the two sentiments, both legitimate, are not necessarily contradictory. In fact, they may indeed happen simultaneously or, alternatively, based on time and place.

Conclusion: culture and urbanism in a global era

For those interested in the study of culture and urbanism in a global era, there are some lessons to be learned.[3] First, if one believes there is a world culture, it is of a form marked by the management of diversity rather than the replication of uniformity. A world culture is also essentially a culture of dominant groups, in which the persistent diversity of the constituent local culture is often a product of globalization itself. This may imply that any convergence indicators only reflect the self-representation of the dominant particular as a true integration of the city of the South into a world system.

The second lesson involves a relationship between world culture and space as created through the increasing interconnectedness of local and national communities. Social theorist Manuel Castells was indeed on target when he pointed to the rise of 'the spaces of places' versus 'the space of flows' in which organizations are connected by the flows of information whose logic is largely uncontrolled by any specific local society, but whose impact is likely to shape the lives of all these local societies (Castells 1992: 73–78). The impact on urbanism is that cultural experience will likely become less place-rooted and more information-based. Some will argue that identity cannot be placeless and that culture cannot be swallowed up by information flows. Indeed, some identities will remain place-based. However, in the present technology-crazed communication era, experience is increasingly shaped by 'virtually reality' environments rather than actual places.[4] We must come to terms with the effect of this change on urbanism in the twenty-first century. In a recent book, this phenomenon is discussed as the 'End of Tradition' (AlSayyad 2003) or the idea that authenticity as a harbinger for historic value is no longer a reliable measure or criteria for judgment or action. It is important to come to terms with the effect of this change on the forms of cities in the twenty-first century.

Finally, and despite academic preoccupation with globalization, the history of the world demonstrates a movement toward cultural differentiation and not homogenization, in which each individual belongs to many cultures, and people have multiple cultural identities. In this sense, identity is always under construction and in constant evolution. If hybridity is accepted as an inherent constituent of national identity, the enduring urbanism must be accepted as only a reflection of a specific transitional stage in life of any society. Indeed, globalization had made the issues of identity and representation in urbanism very cumbersome and has cast doubt on urbanism's ability to represent fully the peoples, nations, and cultures within which it exists. Nevertheless, since culture has become increasingly placeless, urbanism will become its attempts to mediate global domination.

Chapter 12

Space of representational negotiations

Beyond Ruskin and Bhabha[1]

Andrzej Piotrowski

Introduction

Although buildings are commonly perceived as the most permanent material manifestations of culture-specific ideas, their symbolic functioning is anything but unequivocal. The traditional history of architecture has created the impression that built environments transmit well-defined meanings across time by filtering our view of supposed monuments of architecture according to the way they representationally affirm dominant narratives. I argue, however, that many constructed spaces that do not project conclusive messages have nonetheless played an instrumental role in shaping cultural identity, supporting highly nuanced nonverbal cultural exchanges by engendering original thoughts, directing imagination, and focusing critical attention. A historical chapel from Poland will exemplify how the symbolic constitution of such a material place reaches beyond both traditional and contemporary understandings of cultural identity, the first represented by Ruskin and the second by Homi Bhabha.

Epistemology of architecture and the nineteenth century's legacy

The aspects of architecture I discuss here have been largely overlooked because, as mentioned above, buildings have been studied primarily to show how they support politically biased interpretations of history. Much as the cultural identities of colonized peoples have been defined by their colonizers, the best known achievements of architectural creativity have been described, photographed, and discussed in ways that affirm tacit assumptions that support traditional historical periodizations or taxonomies of cultural difference. In such a view, any symbolically significant monument reinforces the accepted understanding of why certain nations have dominated others and how their epistemological and economic concepts have defined those others as inferior and less self-aware. It is not surprising that questions about identity and its material expressions have arisen with an unprecedented sense of urgency as it becomes increasingly apparent that old narratives of cultural singularities and essentialism are inadequate to explain phenomena of the postcolonial and globalized world. While the notion of long-entrenched national identities retains its appeal for many, I would contend that those concerned about their loss in the dynamically changing and hybrid world in which we are currently living are expressing a subconscious longing for the comfort of easy distinctions and judgments inherited from the nineteenth century.

The nineteenth century was quintessential both for the way in which it sorted out cultural identities and the way in which architecture was integrated into these epistemological mechanisms. During this period, the colonial powers developed a broad spectrum of representational devices used to reorganize how people perceived others and themselves. Cultural differences manifesting themselves in appearance, habits, or language were epistemologically mapped out and even artificially created and displayed to support the new world order and its taxonomies (see, e.g., McCrone 1992 or Poignant 2004). New branches of knowledge, such as ethnography, were created to sanction these processes. The nineteenth-century obsession with history is symptomatic of the many ways in which elites constructed the impression that the dominance of certain civilizations was a natural result of the accumulated and consistent efforts of superior individuals and communities. The complex processes of establishing historical facts and their interconnections developed during the period involved both identifying undeniable evidence of past events and dismissing other, less politically useful information. In many cases, it was not simply the sequence of events but historians' and politicians' ways of selecting and representing those events that produced what came to be accepted as continuities in history or cultural integrity. Many nation-states, such as Germany, Italy, and India, emerged in the form we know today by becoming homogenized in this way – assembled out of various degrees of cultural difference.

Architecture also played an instrumental role in these processes of political construction. John Ruskin provides perhaps the best example of how nineteenth-century theoreticians viewed the constitution of identity and its expression in built environments. In his *Seven Lamps*, for example, Ruskin equates the sense of national identity with the memory and awareness of the nation's glorious past, asserts that poetry and especially architecture provide the best mnemonic devices for embodying and preserving such symbolic signs, and argues that it is incumbent upon designers to actively shape and refine this sense of national identity (Ruskin 1989: 178). The tacit assumptions behind these claims, however, are revealed in his choice to explain the desired functioning of such constructions in the chapter devoted not to memory but to obedience. There he asserts that a national style should be constituted like language so it could be taught, internalized, and reproduced like rules of grammar and vocabulary. The degree to which this style is invented or imported is less important to Ruskin than the way it facilitates control over the conceptualization and appreciation of what he considers symbolically proper architecture. In his own words, the goal of establishing a national style is to ensure that 'no individual caprice shall dispense with, or materially vary, accepted types and customary decorations' and 'every member and feature of the architecture of the nation shall be as commonly current, as frankly accepted, as its language or its coin' (Ruskin 1989: 202). A national style, the embodiment of national identity, thus must not only be consistent and pure but provide a perfect didactic tool for eliminating symbolic vagueness and disciplining common perceptions of the structures of belonging. Ruskin deemed architecture instrumental in these processes because, as an aesthetic practice, unlike verbal communication alone, it could tacitly permeate everyday life and shift viewers' attention away from the effort to understand to uncritical appreciation. Even today, reliance on these nineteenth-century assumptions regarding the need for and the routines of a national uniformity promotes a reductive tendency in the study of historical cultures' material expression and of the power–knowledge relationships that shape our perception of them, thus prolonging those tacit processes that Edward Said (1993) uncovers in the history of colonized cultures. By promulgating a politically charged essentialism, historians and theoreticians have trained people to dismiss anything in architecture that escapes well-established symbolic taxonomies, including forms that are culturally nascent or hybrid.

The Holy Trinity Chapel in medieval times

To help correct the effects of this legacy on our current understanding of architecture and its relationship with cultural phenomena, I here examine one such overlooked building to demonstrate that material production has been much more dynamic and complex than the nineteenth-century model of knowing allows. As this example will show, symbolic exchanges between different cultures or communities have always been far more indeterminate and nimble than those registered by the traditional epistemology described above. Indeed, I will argue, architecture played a key role in these processes not because it preserved well-defined signs of symbolic agreements but rather because it admitted and revealed conflicts and contradictions. While the canon of architectural history elaborates only on cases in which the dominant powers used architecture to stabilize and disseminate their ideology, the world is actually saturated with historical and contemporary environments in which buildings perform representational negotiations. Such symbolically dynamic compositions are in fact ubiquitous because architecture has always (and everywhere) given form to culturally nascent thoughts often before words could contain and stabilize them.

To help substantiate this argument, I have chosen Holy Trinity Chapel (Kaplica Świętej Trójcy) at the Royal Castle in Lublin, Poland,[2] in large part because it is a Gothic structure. As such, according to Ruskin's argument, it is the very kind of architecture essential to distilling national distinctions in places such as England, Venice, or France (1989: 208) and symbolically integrating a political domain around the master narrative of an ideologically conscious and deliberately acting ruler. The chapel, although masterfully decorated, is little known abroad because, like the great majority of historical buildings in the border between what was then the Kingdom of Poland and the Grand Duchy of Lithuania – especially those built before the eighteenth century – it seems stylistically imperfect. Architectural historians often call buildings of this kind provincial because their formal features are neither stylistically pure nor consistent and thus do not neatly fit within the dominant taxonomies of architectural periodization or styles.

The geographical location of Poland undoubtedly contributed to this seeming incoherence. The eastern edge of Poland, the border of the Latin West since the Middle Ages, consisted mostly of wooded lowland, not only exposing the country to military invasions but also encouraging cultural and commercial exchanges with its eastern neighbors. From the very beginning of Polish history, religious and political alliances played a crucial role in establishing a military equilibrium across that permeable border. In 988 CE, the conversion of Vladimir the Great, the ruler of Kievan Rus, to the Orthodox religion (a decision that consequently made the Russian Empire the heir of the Byzantine glory and ambition) was prompted by the expansion of Latin domination to Eastern Europe, the conversion of Poland to Catholicism in particular. Until approximately 1500 CE, the territories east of the Wisła River were scarcely populated and underdeveloped. Even after the major political players pledged allegiance to one of the two competing religious powers of Europe, the Kingdom of Poland to Rome and the Kievan Rus with the Duchy of Halicz-Vladimir to Constantinople, the eastern periphery of Poland and the Grand Duchy of Lithuania created a gray zone where the leaders of smaller population centers chose and/or changed religious affiliations depending on the current political situation. This was also possible because in many areas the common people had never abandoned their ancient rites; according to Jerzy Kłoczowski (2000: 55), in the fourteenth century the Grand Duchy of Lithuania was the last mostly pagan state in Europe.[3] Even this late into the religious polarization between Eastern and Western Christendom, and despite their strong family ties to Orthodox Christianity, the rulers of the Duchy remained pagan and treated the prospect of conversion to either branch of Christianity as a political tool useful primarily to gain or restructure power relations in this part of Europe. Jogaila (Jagiełło), one of the brightest politicians leading this multi-religious state, established strong ties with the eastern provinces of Poland and managed to convince other members of the ruling elite that the Kingdom of Poland and the Grand Duchy of Lithuania were compatible enough to function under one ruler. In 1386 CE, Ladislaus (the Latin name given to him at his Christian baptism) Jogaila was crowned the king of Poland as Władysław Jagiełło and thus established a new dynasty and a personal union between the two states, a form of rule in which two or more states have the same ruler but maintain their own boundaries, laws, and identities. The citizens of Lublin, which is located approximately halfway between the two capitals, Kraków and Vilnius (Vilna, Wilno), were instrumental in setting up the political alliance.[4]

The town was also the site of another kind of negotiation, a symbolic engagement between Roman Catholicism and Eastern Orthodoxy. Although the two religious systems were still in political conflict, in that part of Europe the coexistence of people who were separated ideologically by the Great Schism of 1054 CE was a fact of everyday life. The union between the two states was defined by these difficult religious issues, and Władysław Jagiełło designated the newly completed Holy Trinity Chapel at the Royal Castle in Lublin to be a site of such a representational encounter.[5]

The Holy Trinity Chapel is a small structure consisting of two interior volumes stacked vertically, the main hall above and the crypt below. Its exterior is relatively simple and explicitly Gothic in its proportions and its construction methods, including the use of pointed arches for openings and brick as the primary building material. In contrast to the Western-style exterior, however, the interior decorations are Russian-Byzantine, commissioned by Władysław Jagiełło and finished in 1418. Figure 12.1 shows the multiplicity of paintings that cover the walls of that relatively small volume of space. Although unperceivable in the black-and-white reproduction here, color intensifies the viewers' experience of being surrounded by a tapestry of religious icons. As in many

e 12.1 The Holy y Chapel in Lublin, or.

Orthodox interiors of churches in Byzantine or post-Byzantine Greece, the Balkans, or Russia, in Lublin, the viewers' vision and imagination are captured by evocative imagery. Figurative compositions cover all the walls, floating against the bluish background. Lines painted on vertical surfaces create areas for individual images, which only loosely relate to such Gothic features as the shapes of walls or the placement of windows and doors, and the multiplicity and arrangement of paintings appear to defy the physicality of the walls. Nonetheless, as seen in Figure 12.1, the Gothic elements of the structure are also strongly visible as the ribs of the vaulted ceiling collect the forces of gravity and distribute them to the walls. As I discuss elsewhere (Piotrowski 2011: 1–32), the Gothic treatment of structure was one of the most explicit departures from the Byzantine attitude toward architecture. In a Gothic cathedral, unlike the materiality-denying interiors of Byzantium, the material structure manifested an ideal order – a rational perfection, the City of God recreated on Earth. In Lublin, however, the two kinds of symbolism that had separated the Christian East and West coexist again. Although the structural ribs are covered with colorful patterns, such added decoration is not meant to diminish their materiality and blend them with the bluish background, as in a Byzantine space, but remain strongly discernible and visually function like lines painted on walls, in effect organizing the vaults just like the flat borders divide the vertical surfaces.

Like other scholars writing about the chapel (e.g., Walicki 1930; Buczkowa 1983; Durakiewicz 1995), Anna Różycka-Bryzek, an expert on the chapel's paintings, concludes that while their system of display and the symbolic program are of Byzantine origin, their execution was deeply compromised by its incompatibility with the Gothic building (2000: 30, 35). However, if we were to approach this issue not as a matter of unfortunate conflict-producing stylistic imperfection but rather as a representational experiment intended to test a hybrid concept of religious space, other features of the chapel become more relevant. In Lublin, the agreements between the Western and Eastern symbolic systems become at least as meaningful as the articulation of their differences. If we consider the Gothic and Byzantine legacies not as sets of rules but rather as ways of thinking, we can see the strongest tension between them in the center of the chapel.

As seen in Figure 12.2, a plan of the chapel, the building consists of the main hall and the presbytery attached to its east side. The dotted lines indicate the system of ribs on the vaulted ceiling. While Byzantine churches were primarily composed around the central empty space of the naos and its vertical axis, Gothic churches emphasized a horizontal procession towards the main alter. These differing layouts assume a different sense of symbolic directionality, which was reinforced by different distributions of daylight. In the chapel, only the presbytery implies the horizontal directionality of a Gothic space, while the footprint of the main hall is square and thus similar to Byzantine and post-Byzantine churches. According to Robert M. Kungel (2000: 37–54), the Gothic style was also used in Orthodox churches on the western outskirts of Lithuania in the fifteenth and sixteenth centuries. Often their size and layout were similar to that of the Holy Trinity Chapel, with one exception: their central space was structured by

ure 12.2 The Holy
nity Chapel in Lublin,
n.

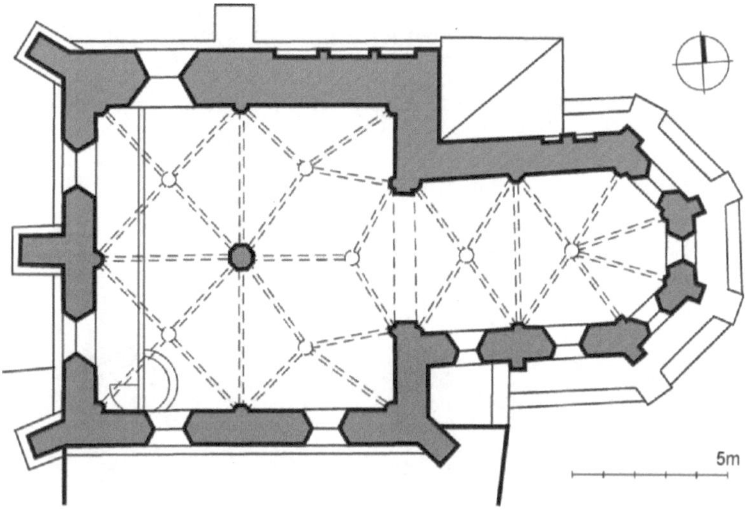

four columns, leaving the very center of the church empty, an arrangement characteristic of the Russian Orthodox church model, where the central space is taller and infused with light coming through windows of a large lantern above. That experiential quality of light captured at the top of the church interior was symbolically crucial for the Eastern models of religious representation (Piotrowski 2011: 1–32).

In contrast, Figure 12.3 shows that in the Holy Trinity Chapel the center space is occupied by the most material and technically explicit of the building's elements – a column revealing how human ingenuity has transferred the forces of gravity from the ceiling to the ground. This representational shift redirects thoughts away from the Byzantine vertical progression of symbolic hierarchy and towards the Western symbolic progression to the altar as the symbolic center.[6] This is a radical juxtaposition of two modalities of religious thought: in the symbolic center of a Byzantine space of visual representation an ephemeral phenomenon is replaced by a material statement of rational ingenuity.

According to the standard taxonomy of historical styles in Western architecture, an interior like this should not exist, although it does. That does not mean, however, that the Holy Trinity Chapel is a compromised or imperfect space of representation. Nor was its juxtaposition of the Byzantine and Latin modes of symbolic ordering meant to dominate or subvert the principles of either belief or identity. Although the chapel paintings were probably commissioned right after one of the most significant military victories of Jagiełło's reign, the defeat of the Teutonic Knights and their Western allies under Grunwald (Tannenberg) in 1410 that practically ended religiously motivated attempts to colonize the northern parts of the Grand Duchy of Lithuania by the Western powers, the symbolic arrangement of this place of worship does not seem to take political sides. Rather, the chapel is a material manifestation of thought seeking new

Figure 12.3 The Holy Trinity Chapel in Lublin interior.

ways to identify and disarm schismatic divisions within European Christianity. This place of worship reflects a way of thinking that is not ignorant of differences but rather refuses to follow the centuries-old patterns of constructing ideological conflict for political reasons. All the relevant principles of religious thought from these traditions are acknowledged, but none of them is allowed to dominate the entire spectrum of visual experiences in the chapel. Nor does this space of representation manifest a new religious program, although it may not be a coincidence that the paintings were completed only 21 years before the Union of Florence (1439 CE), the most successful attempt to reconcile Constantinople and Rome.[7] Just as the personal union between the two states was not meant to homogenize its people, the chapel was not intended to suppress their cultural differences but was designed to explore the symbolic interactions between distinct beliefs and traditions. What might have been a common practice of people living together on the border between Poland and the Grand Duchy of Lithuania was represented here as a religious material environment searching for a new symbolic and political order.

The chapel in the Reformation period

Nor does the history of this building as the manifestation of complex processes involving cultural and religious identity end at the time those paintings were added at the beginning of the fifteenth century. In the sixteenth and seventeenth centuries, as we can see in Figure 12.4, late Renaissance decorations were added to the chapel's Gothic front façade.

Around 1521 CE the elevated doorway was embellished with decorations carved in white sandstone, and a century later an elaborate composition was superimposed on the roof gable,[8] new design patterns that have nothing in common with the Gothic articulation of the building. Their features are also

ure 12.4 The Holy ity Chapel in Lublin, erior.

unorthodox in relation to the stylistic systems promoted by the primarily Catholic political establishment of the time. The slight misalignment of the arch and its supporting pilasters and the strangely inflated capitals in the doorway frame, for example, do not conform to the Roman aesthetic principles that were influencing the ruling elite in Poland at that time. The new composition of the gable consists of different forms, including spirals, short obelisks, and the layers of shallow arches and pilasters. Independent of the Gothic order below, it produces a highly complex skyline that adds an almost playful character to this otherwise austere Gothic exterior. Altogether, these later additions seem totally oblivious of the design logic that preceded them. They create a tension with the older structure that resembles a sensibility similar to Mannerism, a spectrum of design practices characteristic of places in Europe where architecture and the visual arts were used to test the limits of Roman domination (Piotrowski 2011: 99–131).[9] Even Italy has many similar examples built in places and at times in which people rebelled, according to Salvatore Caponetto (1999), against the political powers behind the Catholic ideology.

I would argue that, like the conflicted relationship between the Byzantine paintings and the Gothic shell of the building, this seeming mismatch of exterior elements actually represents profound and unique cultural processes that took place in Poland and Lithuania in the sixteenth and seventeenth centuries. The two countries, initially united politically by Jogaila, grew accustomed to accepting cultural and religious differences. In addition to the mixture of Poles, Ruthenians (today's Ukrainians), Belorussians, Russians, and Lithuanians who always inhabited the eastern peripheries of Poland, many Jews who were persecuted in other countries also found refuge in this part of Europe. Smaller groups representing other cultures and religions, including Karaites accepting only the Old Testament, Monophysitic Armenians, or Islamic Lithuanian Tartars, settled on these territories as well. Yet the biggest influx of people to the Commonwealth of Poland and Lithuania happened in the sixteenth and seventeenth centuries, when Lublin also became home to Italians, Armenians, Germans, Scots, and Magyars,[10] some of whom migrated for economic reasons but many of whom were escaping religious persecution or the religious upheaval triggered in Europe by the Reformation and the Counter-Reformation. The Lublin region attracted some Lutherans and many Calvinists with a progressive sub-group of *Arianie* (Arians, also known as *Bracia Polscy*, Polish Brethren, or *Socynianie*, Socinians).[11] The arrival of these Protestant immigrants also triggered a wave of conversions in Poland and Lithuania. Consequently, the Commonwealth became what Papal officials called an *asylum hereticorum* (heretics' safe haven).[12] This is the symbolic environment in a state of symbolic uncertainty, contested by some but actively redefined by others, that the Holy Trinity Chapel represents. An intellectually vibrant community of people seeking a new tolerant order used architecture to test these ideas, searching for a political system accommodating diversity like the added pieces on the chapel's façade. While also fostering some of the most progressive laws in Europe, such as the Compact of Warsaw, they used architecture to explore hybrid notions of cultural and religious identity.[13]

Identity and the logo-centric turn

Many important buildings embodying similar efforts still exist in the territories of the old Commonwealth of the Kingdom of Poland and the Grand Duchy of Lithuania. Similar examples of people using material environments to test those sense-making processes to explore who they are and how they interrelate can be found all over the world, and today's global economy portends that their production will increase exponentially. That we currently know so little about them reveals the magnitude of the epistemological problem reflected in Ruskin's ideas and still operating when we dismiss the Holy Trinity Chapel as a charming but provincial aberration. Ruskin's notion that all cultural production is conscious, linguistically controllable, and validated by the integrity of its narrative is also deeply built into contemporary cultural theories. When one assumes with Gayatri Chakravorty Spivak, for instance, that 'we know no world that is not organized as a language, we operate with no other consciousness but one structured as a language' (1987: 77–78), it is difficult to imagine that people may produce architecture not to communicate already well-formed messages but to expand imagination and to think new thoughts. Paradoxically, some of the most advanced ideas emerging from recent postcolonial theories are limited by similar problems. Consider Bhabha's Third Space of articulation, a new kind of interstitial space of symbolic exchanges 'where the negotiation of incommensurable differences creates a tension' (Bhabha 1994: 218). While moving beyond the simple notion that the knowledge of culture is that of an integrated code of signs and references and beyond binary oppositions and universal categories, Bhabha's concept still engages cultures primarily as linguistic constructs (Rutherford 1990: 210). Intended to represent a mental environment in which different people maintain a dialogue by accepting the hybridity of their cultural condition, the Third Space, by Bhabha's own admission, is 'unrepresentable in itself' (Bhabha 1994: 37).

In many respects, materially constructed spaces may function similarly to Bhabha's conceptual Third Space of articulation, as both inform identity by exposing people to manifestations of cultural issues and tensions among them. A significant difference between the two, however, is that architecture gives priority to the processes of shaping thought while the Third Space gives priority to the transmission of messages. Material environments test perceptual manifestations of abstract orders that one mentally inhabits – symbolically relates to, or uses to structure cognitive processes. As the Lublin example suggests, visual environments may both formally reconcile different symbolic systems and help identify irreconcilable differences or gray areas of potential negotiations. This is possible because architecture acts not only on well-defined symbolic content but also – and primarily – on modes of thinking and perceiving. Consequently, built environments engender the imaginative consideration of possibilities outside existing symbolic systems and help envision other structures of thought. Bhabha's concept, on the other hand, gives priority to the verbal understanding of the world and the self, focusing on the difficulties in communication and translation because his linguistic model assumes that verbal clarity of definitions and interpretations is the ultimate goal of cultural interactions.

As the Holy Trinity Chapel exemplifies, buildings have long negotiated cultural differences and tested new ordering of symbolic reality in ways reaching beyond verbal discourses. Not just mere illustrations of narratives, physical structures have frequently operated on these nascent thoughts that determine what aspects of dynamically changing reality become perceivable, culturally relevant, or altogether thinkable. By admitting complex ideas, revealing their difficult interactions, and exposing them to many people over extended periods of time, architecture has helped communities share explorations of important but frequently inconclusive thoughts. Spaces of representation, such as that of the Holy Trinity Chapel, question conventional expectations and highlight tensions within emerging symbolic orders before words can reconcile and stabilize them as ideological or political concepts. As some literary scholars of identity have noted (Moya and Hames-Garcia 2000 and especially Mohanty 2000), the spectrum of thoughts that shape one's ability to perceive, imagine, or remember, and how these abilities relate to our emotional attitudes or subconscious biases, remain the most insufficiently explored of cultural issues. In this respect, architecture provides a greatly overlooked and undervalued source of information. What is needed is a new approach to studying the relationship between architecture and cultural identity that would include studying currently marginalized spatial practices with a new kind of scholarly precision. It should be possible to explore both contemporary everyday environments and historical monuments in a way that combines the non-verbal mode of design thinking with cultural theories, creating an epistemological emphasis that will not only complement current studies but radically expand our understanding of complex and location-specific cultural issues.

Chapter 13

Colour and tradition

The portrayal of Mexican architecture in the American press

Catherine Ettinger

Introduction

At the outset of the twentieth century, as Mexico emerged from a civil war,[1] the definition of the essence of Mexican identity was a topic central to intellectual discourse. A revision of the discussions among architects during the decades following the war – the 1920s, 1930s and 1940s – reveals a perspective in which architecture was to play a double role, solving social demands while at the same time participating in the definition of Mexican identity.

In this quest for identity, the first generation of post-revolutionary architects debated whether the indigenous or the Spanish colonial were the legitimate roots to which to return. Manuel Amábilis, in the Yucatán peninsula, favoured the first, expressing his ideas through his neo-Mayan buildings and defending his position in terms of an ontological relationship to place (Ríos Garza *et al.* 2001: 42; Amábilis 2003: 34). Jesús T. Acevedo, Nicolás and Federico Mariscal, as well as many others, proposed the neo-colonial style as authentically Mexican, since it absorbed the experience of the Mexican baroque, in itself *mestizo*[2] or hybrid. While the debate between the virtues of neo-indigenism and neo-colonialism was being waged,[3] the modern movement was making its appearance in the country. Promoted by one group of talented architects, modernity was viewed as foreign by others, as if it were another experience in copying the architecture of distant countries while ignoring the singularities of Mexico, its land and its past. The divergent positions were expressed in an event organized by the Society of Mexican Architects (SAM) in 1933. In the series of talks that was subsequently published, concern about the relation between international trends and the creation of a Mexican architecture was ever present (Ríos Garza *et al.* 2001).

The discussions engendered during the 1920s and 1930s evolved over the century, but the central issue of defining what is Mexican architecture and its relation to that which is foreign continued to be problematic. Under the guises of discussions on regionalism and globalization, the debate is present even today. One of the aspects to be noted in this discussion is the omnipresence of that which is foreign and of the viewpoint of the foreigner. Arguments presented in favour of, or against, any of the three trends often called upon the authority of the foreign observer or disdained the imitation of foreign architecture.[4]

With this in mind, this chapter explores the viewpoint of the foreign press in reference to Mexican architecture. In the literature reviewed, by the mid-century an image of Mexican architecture as a colourful reinterpretation of popular art and rural traditions became increasingly present. By the 1980s, the diversity existent in Mexico and the importance of international architecture were rendered almost invisible as Mexican architecture was reduced to the image created by Barragán, his followers and large hotel chains. This text illustrates the ways in which the foreign press participated in the consolidation of this image and its appropriation by Mexicans as a representation of identity.

Mexico in the American imagination

During the first half of the twentieth century Mexico loomed large in the American imagination. Several factors were at work in the flowering of cultural relations between Mexico and the United States. Starting in the 1920s, as Mexico entered a period of political stability, transit between the two countries increased. In 1921 Mexico, anxious to promote a new image outside of the country, opened a summer school for foreigners that offered courses in Mexican culture and literature with the participation of notable Mexican intellectuals (Delpar 1992: 19). At the same time, in the United States, the presence of Mexican artists and intellectuals such as the muralists David Siquieros and Diego Rivera contributed to greater knowledge and a better understanding of Mexico (Delpar 1992: 41).

During the 1920s and 30s Mexico attracted a great number of foreign residents, many of them disenchanted with modernity, others sympathizers of the Mexican revolution. Journalists such as Carleton Beals and Frank Tannenbaum, the anthropologist Anita Brenner, artists and architects including René d'Harnoncourt from Vienna, William Spratling from the United States and writers including Katherine Anne Porter, D.H. Lawrence and Graham Greene formed an important expat group linked closely to the intellectual elite in Mexico City. Through their stories and novels set in Mexico they created an image of a vibrant life in a traditional setting. Another American woman influential in the foreign press was Frances Toor who edited the magazine *Mexican Folkways*, which promoted tourism as well as Mexican arts and crafts among English speakers.

In this literature, architecture was present as a backdrop, a scenery of intense sensuality. Colour, lush vegetation, and adobe ruins appear in a sort of primitive romanticism that emphasized the beauty of traditional life. Books such as *Mexicana: A Book of Pictures*, published by René d'Harnoncourt (1931) or *Little Mexico* by William Spratling (1932) strengthened this image as did literature set in the country.

Through their work these foreigners endorsed a romantic image of Mexico abroad in which modernity and tradition were placed in contraposition. There were recurrent topics in this literature including a distinct conception of time in Mexico, the spiritual superiority of the native and the value of handicrafts and manual labour in general. Houses and towns are described with attention to irregularity and a setting in tropical vegetation. In this portrayal modernity looms as a threat and comparisons with the United States emphasize the richness of life in Mexico in contrast to the impoverished modernity of industrialized society. A case in point was the monograph written by Elizabeth Morrow (1932), wife of Dwight Morrow, the ambassador to Mexico from 1927 to 1930 to describe the couple's house in Cuernavaca. The name given to the house and the book, *Casa Mañana* ("House of Tomorrow"), echoes one of the constant themes in the foreigners' view of Mexico, namely the vagueness of references in time. Morrow's text describes with great detail the experience of building in Mexico and her dealings with workers as well as the beauty of irregularity and the handmade. This book complemented the exotic portrait of life in Mexico given by contemporary authors and set the stage for future portraits of Mexico's architecture.

Mexican architecture in the Mexican press

Several magazines served as national media for promoting a modern image of Mexican architecture in the twentieth century. The earliest, *El arquitecto*, was published from 1923 to 1927 (Ríos Garza 2004a: 7) by the Sociedad de Arquitectos Mexicanos, an organization formed in 1919 when a group of architects split off from the former gremial organization that included engineers. As a result, part of the discourse was dedicated to the recognition of architecture as an art, in contrast to engineering as technique. In spite of this, the magazine published articles on modern architecture but tended to present the styles that were contributing to the creation of a national culture, such as the neo-colonial and neo-indigenous.

The most relevant example of the architectural press in twentieth-century Mexico was the magazine *Arquitectura /México*, with a total of 119 issues published between 1938 and 1978 under the direction of Mario Pani (Ríos Garza and Noelle 2008). With a modern architect such as Pani at the head of the magazine it is not surprising that it promoted almost exclusively the works of modern architects including Mario Pani, José Villagrán García, Juan Sordo Madaleno, Jaime Ortiz Monasterio, Pedro Ramírez Vázquez and Enrique Yáñez with a great deal of attention given to projects financed by the state. There were special issues dedicated to schools, hospitals, markets, factories and UNAM's campus, *Ciudad Universitaria*. José Villagrán wrote regularly for the magazine between 1939 and 1943 presenting theoretical concepts he was teaching at the time, as did Vladimir Kaspé, a Russian immigrant who was influential in introducing modern architecture in Mexico. It is noteworthy that Luis Barragán is mentioned only three times in the forty years the magazine was current. In contrast to what we will observe in the foreign press, the mention of tradition rarely appears, although the notion of Mexican-ness in architecture is present. Even in the case of the counter-establishment publication

Arquitectura y lo demás which appeared from 1945 to 1950 (Ríos Garza 2004b), the emphasis was on modernity.

Mexican architecture in the English language press, 1927–1960

As the Mexican press constructed an image of modernity, another more traditional image of Mexican architecture was being forged from without. In 1927 William Spratling published two articles on Mexico in *Architectural Forum* (Spratling 1927: 1–8 and 161–168) and two years later another in *Architecture* (Spratling 1929: 75–144). In all three cases they contained a series of freehand drawings that emphasized irregularity, natural materials and tropical plants. In 1931 the *The Architectural Record* published an article on Luis Barragán that showed one of his early houses in Guadalajara done in a style similar to California mission with Moorish details (*The Architectural Record* 1931: 62–64). *House & Garden* (1931: 80, 56–57) also published a piece on Barragán the following month and another titled 'Modern in Mexico' (1937: 72, 40–41). In 1935 two articles on Barragán appeared in *The Architectural Record*, one on a children's park in Guadalajara and another showing his recent work (*The Architectural Record* 1935a: 77, 33–46 and 1935b: 78, 165–169).

In 1936 the first article on Mexico that showed international style architecture was published in *The Architectural Record* (Noelle 2004: 52); it showed a boarding school for children built in Mexico City by José Villagrán and Enrique de la Mora (*The Architectural Record* 1936: 79, 445–450); Villagrán was a Mexican rationalist architect and theoretician who was a pioneer in the innovation of the design of hospitals during the 1920s and 30s. The two-paragraph note that accompanied the photographs and plan of the building emphasized the modernity of the institution. In October 1944, Villagran's Children's Hospital in Mexico City was also presented in the *The Architectural Record* (1944: 96, 60–67).

In 1937 Esther Born, an architect and photographer, published an extensive article on Mexican architecture in *The Architectural Record*, and the same material was published in book form entitled *The New Architecture in Mexico* (1937). Born had spent ten months in Mexico, conducting research, photographing architecture and interviewing practitioners. The book features articles by Mexican architects and urbanists, Born's photography with descriptions of projects and short biographies of architects as well as an introduction written by her that expresses her surprise at the quantity of modern architecture to be found in Mexico. She stated:

> Such a quantity would be unexpected in any North American city; but to the Northerner, acquainted with Mexico only through literature and hearsay, the energy displayed and the up-to-the-minute quality are doubly astonishing. We had thought of our neighbours as engaged in pursuits different than ours. These people were our opposites. Their territory was all mountainous, contrasted with our level central

> basin; it was occupied chiefly by Indians, not white men; colonized by Spaniards instead of Englishmen; [. . .] the inhabitants, we were led to believe, supported themselves chiefly by handicraft, lacked a sense of time, were of a mystical rather than a practical bent of mind and, in countless other ways, differed from us as much as human beings could; besides, they were much happier [. . .]
>
> (Born 1937: 3)

In this quote Born, in spite of her desire to show Mexico and its architecture as modern, reveals her surprise and so weakens her premise and, while her interest lay in modernity, she emphasized the singularities of Mexican architecture and did not resist the temptation to include in the book photographs of maguey cactus and bullfights, signalling the coexistence of the traditional with the modern.

The colonialist gesture of the American writers was not lost on the Mexicans. Villagrán, a rationalist hardliner, apparently bristled at the attention the international press gave to the traditional aspects of Mexican modern architecture and to the relationship of Mexican architecture to history. Thirty years later, in a revision of the view of foreigners on Mexican architecture, he stated 'Obviously, Mrs Born did not expect to find, in Mexico, architecture so profuse or so frankly Western in character (Villagrán 1967: 9).

In the 1950s two important publications presented Mexican architecture to an English speaking audience: *Latin American Architecture since 1945* by Henry-Russell Hitchcock (1955) and *Mexico's Modern Architecture* by I. E. Myers (1952). Hitchcock's book was the direct result of an exhibit on modern architecture in Latin America organized by the Museum of Modern Art (MOMA) in New York City. From the start, Hitchcock emphasized the diversity present in Latin America and the difficulties of comparison; at the same time he identified the importance of pre-Hispanic cultures in the gestation of modern architecture, stating that 'particularly in Mexico and in the Andean highlands great prehistoric cultures have left behind monuments comparable to those of Egypt or Mesopotamia.' He noted the particular importance of indigenous heritage in Mexico in describing the *frontón* courts at *Ciudad Universitaria* (the university campus) that 'imitate Aztec shapes' (Hitchcock 1955: 13). Hitchcock described *Ciudad Universitaria* as 'the most spectacular extra-urban architectural entity of the North American continent' remarking on its 'bold color, architecturally scaled external mosaics, and a somewhat insensitive handling of the conventional elements of modern design, [that] achieve a peculiarly Mexican intensity in which references to the Indian heritage, generally pictorial rather than architectural, play a positive role' (Hitchcock 1955: 44).

Hitchcock's choice of words –'insensitive' and 'pictorial'– was probably not well received, in spite of the phrase 'a positive role'. With reference to domestic architecture, Hitchcock illustrated 'the semi-oriental seclusion of Iberian tradition' with the work of Luis Barragán in the houses of the Pedregal (Hitchcock 1955: 45).

Hitchcock's book, due both to his prestige and to MOMA's importance in the art world, was one of the most widely recognized portrayals of Latin American architecture and it circulated not only in English-speaking countries, but was also extensively consulted in Latin America. The book included an introductory text and a series of plates with photographs, plans and concise descriptions of the works of Alberto T. Arai, Luis Barragán, Félix Candela, Max Cetto, Enrique de la Mora, Juan O'Gorman, and Mario Pani, among others.

The second book mentioned, *Mexico's Modern Architecture*, was published by the Instituto Nacional de Bellas Artes (INBA) and promoted by Enrique Yáñez, then head of the Architecture department, with the express intention of promoting Mexican modern architecture among English speaking foreigners (Yáñez 1952: 9). It was also the result of an exhibit – this one organized in Mexico City in 1950 – and follows the format used by Hitchcock. The introductory texts deal with cultural heritage, the people, the land, the economy, the city and modern architecture and are followed by plates with the selected works organized by building type.[5] The introduction was written by Richard Neutra, who had visited Mexico in 1937 and 1951 and had established contact with Mexican colleagues. Undoubtedly Neutra's participation in the publication would help the book abroad. Neutra noted, just as his predecessors had, the importance of indigenous precedents in creating modern Mexican architecture stating: 'Mexico is to my mind the most vital native country of the Americas, where the most modern innovations in architecture and the arts are wedded to indigenous trends of an indigenous people' (Neutra 1952: 21). In this characterization we observe the articulation of a complex relationship between modernity and tradition in which the latter provides authenticity to the former.

Myers' selection of works and texts reflect the INBA's desire to portray Mexico as modern and therefore includes exclusively functionalist architecture. The texts that accompany the plates emphasize the modernity of the examples in contrast to contemporary neo-colonial or mission style buildings.[6] Instead of looking to native precedents, Myers explains Mexican architecture in an ontological relationship to the place, although he recognizes the role of the regional traditions in crafts:

> The land has dictated Mexico's architecture, and until modern technics [sic] reach remote rural areas the land will continue to shape the dwellings of the people. The clay, the stone, the flora, are still the basic elements of construction as in centuries past. In the hands of the Mexican people, these basic architectural materials have been used, consciously or not with great aesthetic accomplishment, and their functional character has not been compromised.
>
> (Myers 1952: 36)

Myer's mention of 'the hands of the Mexican people' alludes to craftsmanship and in spite of the express intention of presenting Mexico's modern architecture to outsiders, he includes discussion of the indigenous, as in the reference to a house by Enrique del Moral:

> Contemporary Mexican architects display increasing recognition of the value of popular rural architecture. In this residence there are many indications of the architect's love of the indigenous. The floor planking of differing widths suggests the informality of regional construction, accented by modern furnishing. Tiles set among the rivers stones of the terrace create a pattern in contrast to the smooth expanse of the walls.
>
> (Myers 1952: 100–101)

The literature mentioned attempts to understand Mexican modern architecture as a unique expression of modernity and tradition. Although references to the native and to the pre-Hispanic are constant, the buildings described are contemporary and follow the principles of the modern movement. Tradition and the past are simply explanations for Mexico's singularities. In general, the intention was to show that Mexico has a valid, unique modern architecture grounded in history and tradition, even if in comparison to the Mexican press some architects, such as Luis Barragán, were given greater attention.

Mexico, colour and tradition: from Verna Cook Shipway to Mexican style

Between 1960 and 1970 Verna Cook Shipway and Warren Shipway[7] published a series of five books on Mexican architecture that were widely read, both in Mexico and in the United States. These picture books marked the beginning of a distinct portrayal of Mexican architecture that emphasized an authentic link to the past and to tradition with few references to modernity. The books deal almost exclusively with domestic architecture although a few hotels are also included; priority is given to decorative details over considerations on space and modern architecture, although present, is placed within the domain of the past. These books were followed by publications on both sides of the border that strengthened and codified a new image of Mexican architecture dissociable from vibrant gardens, massive walls, intense colours and rugged textures.

This inaugural series included: *The Mexican House Old and New* (1960), *Mexican Interiors* (1962), *Mexican Homes of Today* (1964), *Decorative Design in Mexican Homes* (1966) and *Houses of Mexico: Origins and Traditions* (1970).[8] These books are in essence photographic essays that emphasized the sensual experience of traditional Mexican architecture. In lieu of chapters, the finished book presents the material in categories of architectural elements such as facades, patio arcades, balconies, shutters, *conchas*, lighting fixtures, open-walled rooms, etc. in no discernible order.

The titles of the first and last books, *The Mexican House Old and New* and *Houses of Mexico: Origins and Traditions*, both allude to the relationship between modernity and history, yet the examples presented by the authors refer principally to historic content. When modern examples appear, there is always a reference to past tradition – in the grille work, courtyard disposition, craftsmanship, colourful gardens, the use of open rooms or the 'oriental' seclusion. Even in the book entitled *Mexican Homes of Today* the modern examples are few and judging from the names of the owners, the expat population would appear to be over-represented. This makes sense since what the authors were presenting was not simply architecture, but rather a way of life materialized in the house. They were displaying an authentic way of living manifest in hand-crafted details, colourful gardens and peaceful seclusion.

The books illustrate how previous formulations of the relationship between modernity and tradition in Mexico were far from resolved. During the 1960s, Mexico continued to represent spiritual authenticity as the antithesis of the modernity and practicality of the United States. The authors were well versed in modern Mexican architecture and their readings had included *18 Homes of Mexican Architects* published by Enrique Yáñez (1951). References to Juan O'Gorman, Mario Pani, Luis Barragán and even to a young Ricardo Legorreta appear in their notes, yet, in their presentation of Mexican architecture, modernity is shown only when it invokes eternal or spiritual values or illustrates the essence of tradition (Verna Cook Shipway Papers, Box 1, Folder 8).

The Shipways' books were the first in an avalanche of publications on Mexican architecture. The same publishing house edited three more books on Mexican architecture following the same perspective: *The Architecture of Mexico. Yesterday and Today* (Beacham 1969), *Tradition of Craftsmanship in Mexican Homes* (O'Gorman 1980) and *Patios and Gardens of Mexico* (O'Gorman 1988).

An exceptional book, in the sense that it takes a more complex perspective, is *Builders in the Sun: Five Mexican Architects* by Clive Bamford Smith (1967) which presented the work of Mario Pani, Mathias Goeritz, Felix Candela, Luis Barragán and Juan O'Gorman.[9] The title alludes to the importance Hitchcock had attached to the sun in his revision of Latin American architecture two decades earlier and the book follows a similar format, with an introduction on each of the architects followed by plates illustrating their work. However, in contrast to other publications on Mexican architecture abroad, Smith's book does not attempt to generalize; instead he presents each architect's work as unique and allows the author's words to permeate the discourse with sensitivity to the local perspective. The selection of architects is diverse; while O'Gorman, Barragán, Candela, and Mathias Goeritz are quoted as critical of functionalism, the book closes with Pani who continued to be committed to functionalism and the possibilities it provided for the solution of social problems, specifically housing. This book presents a more balanced view of contemporary Mexican architecture, which was much more palatable to José Villagrán than other foreign texts, as manifested in the introduction he wrote for the book.

Particularly during the 1980s and 1990s the picture book illustrating Mexican architecture in full colour not only proliferated in English, but also became standard fare on the south side of the border. The first complete book published on Luis Barragán was MOMA's edition by Emilio Ambasz (1976) based on the exhibit of his work. The popularity of Barragán and the publicity that resulted in his being awarded the Pritzker Prize in 1980 most certainly contributed to the interest in Mexican architecture that would follow.

Among the publications to be mentioned is *Casa Mexicana: The Architecture, Design, and Style of Mexico* by Tim Street-Porter (1989), one of the first widely distributed books. This was followed by the *Hacienda* series initiated by Karen Witynski and Joe P. Carr (1999) which now includes a total of eight books that promote an image of Mexican architecture related to country living, history and crafts. The interest in Mexican architecture was picked up by other publishing houses including Chronicle and Schiffer, among others.

As the popularity of this type of book grew, particularly among a foreign audience, the Mexican paint company, COMEX, began its own series, at first focused on colour, albeit with financial gain at the centre of the enterprise through the sale of paint. The company has published over fifteen volumes to date, most of them dedicated to Mexican architecture although in the last few years more attention has been placed on international architects, particularly in the two most recent books. The books initially showed mostly works that followed the tradition of the School of Jalisco with the use of bright colours, regional materials and rough textures, although more recent publications have branched out to include a greater variety of expressions.

The interest in Mexican architecture, both within the country and abroad, gave birth to Arquitectos Mexicanos Editores in 1997.[10] This specialized editor based in Mexico City has published over 50 bilingual books in four main series (*Arquitectos Mexicanos*, *Espacios*, *Interiores Mexicanos*, and *Ideas*). These books emphasize particularly Mexican qualities in materials, colour, and craftsmanship.

To be noted in the editions of COMEX and AM Editores is the way in which what was originally the viewpoint from without, became a viewpoint from within. The series of Mexican publications mirrored a tendency that began in the United States in the portrayal of Mexican architecture almost exclusively in terms of tradition, craftsmanship and colour. This tendency coincided with international interest in regional schools and so the role of the foreign press certainly should not be viewed as the only factor involved.

In conclusion: a play of mirrors

The publication of books composed mostly of photographs is certainly not unique to Mexico; nor is the portrayal of its architecture as exotic. Large publishing houses have edited numerous glimpses into 'non-Western' cultures through architecture and it would appear this trend will continue. For the case of Mexico I would like to point out two aspects. First to be noted is the awareness in Mexico of the role that architecture plays in the construction of national identity, an idea that was present in architectural discourse from the early twentieth century onwards. Secondly, a local sensitivity to the perception of Mexico by outsiders that appears in the discourse of philosophy and of architecture during the first half of the twentieth century. These two aspects are related to the way in which certain traits were exacerbated in the reappropriation of the image of Mexicanness in architecture during the 1980s and 1990s.

With reference to the first point, from the first decades of the twentieth century architecture was deemed central to discussions on Mexican-ness and on the true roots of Mexican culture. There was awareness on the part of architects and intellectuals of the usefulness of architecture as a tool in the construction of a national identity; the campaign carried out by José Vasconcelos as Minister of Public Education for the design of elementary schools in neo-colonial style is a case in point. A process of identification with architecture that began from within Mexico at the end of the Mexican Revolution continued in the decades that followed and was reinforced from sources outside of Mexico. It could be argued that architecture never lost this role in defining identity and continues to be a key element in this process.

Secondly to be noted is an acute sensitivity to the viewpoint of foreigners on the part of Mexicans; this awareness is linked to the construction of Mexican identity throughout the twentieth century and visible in the discourse of philosophers such as Samuel Ramos and Emilio Uranga as well as in that of architectural theorists such as Alberto T. Arai and José Villagrán. At the mid-century Arai (2001) wrote *Caminos para una arquitectura mexicana*, literally *Pathways to a Mexican Architecture*, continuing, through reflections on architecture, an ongoing discussion among Mexican philosophers on the essence of Mexican-ness and its roots in the indigenous or in the Spanish. Arai's text reflects concern for the relationship between the local and the universal, a concern that remained at the centre of discussion throughout the century.

Particularly illustrative of the awareness of the foreign perspective on Mexican architecture is a critical essay written by José Villagrán in 1967 as the prologue to *Builders in the Sun* (Smith 1967). In his discussion of foreign perception, referring specifically to Esther Born and the book *The New Architecture in Mexico*, he stated the following:

> It can be deduced that the mentality of the foreign observer at that time – a mentality which has not changed fundamentally—could not conceive that a country such as ours [. . .] should be in step with what was believed by Europeans to be their so advanced thought, achievement and accomplishment and even should be doing things with a 'different accent'.
>
> (Villagrán 1967: 9)

Villagrán also commented on the Pan American Congress of Architects that had taken place in Mexico City in 1952 expressing the belief that modern architects from outside of Mexico had been unable to understand what Mexican architects were attempting to do. Referring to the delegates he stated:

> Some of them, confronted by basalt walls of considerable thickness and height, and pavements of the same material, the rough surfaces of which differ so greatly from those of the granite or polished marble floors which the European and North American are accustomed

> to tread, were unable to conceal their astonishment and revulsion at such manifestations of 'barbarity'—the only term they could find to describe such treatment of basalt [. . .] Many others shared these opinions, and it would be interesting to record their names now that, in their work and their criticism, they extol the quality of what they then called 'orgies of colour, incomprehensible to any sensible Western architect'.
>
> (Villagrán 1967: 9)

While Villagrán admitted that Mexican architecture had characteristics that remained present over the centuries, he also defended it as mainly Western, criticizing those who considered that it had lost its essence as it entered modernity. Villagrán's acute awareness of the portrayal of Mexican architecture as native, traditional and rooted in the past provoked irritation, as he attempted to present the work of Mexico's architects as primarily Western, advanced and modern.

Perhaps Villagrán foresaw the success the image of the traditional and the colourful would have in the transmission of Mexican architecture, an image, born of the idealization of tradition on the part of foreigners, which was reappropriated in Mexico during the 1980s. The process of reappropriation coincided with discussions on the regional, particularly critical regionalism, in architectural schools all over the world. In México, the regional became synonymous with the School of Jalisco and Luis Barragán and to a lesser degree with the works of Abraham Zabludovsky and Teodoro González de León, who also explored the possibilities of regional expressions through references to pre-Hispanic architecture and the use of rough surfaces.

The image of Mexican architecture as colour and tradition is not limited to the professional sphere, but rather has permeated diverse social groups in Mexico. Many subdivisions require stylistic conventions in houses built within their confines; often cited as 'contemporary Mexican' or 'rustic', these conventions would appear to refer to the image mentioned, but also often include the use of decorative elements such as stone arches, balconies and iron work.

It would seem we have a case that fits Edward Said's schemes, in which Mexican architecture conforms to the image created from without. The applicability of postcolonial theory to Latin America has been questioned; yet the relationship of the region's countries to former colonial powers and to the United States has made it useful for the explanation of a variety of phenomena (Moraña *et al.* 2008: 1–22). In the case at hand, the texts themselves provide credence for this idea in the way that they lend authority to the viewpoint of the foreigner. Lozoya (2010: 188) considers that the continuity of identitary forms related to regional architecture and the notion of tradition is due, in part, to awareness on the part of Mexican architects of belonging to a periphery, of lacking cultural leadership and of participating in modernity in a different temporality. This being so, the attention given to the perspective of the foreign press makes sense. At the same time it is important to recognize that there are other factors at play such as the association between traditional imagery and the domestic sphere and the popular taste for familiar forms over modern abstraction. In no sense could the strength of the image of Mexican architecture be considered a creation of the foreign but rather a product of reflections and exchange over nearly a century that sheds light on issues of architecture and identity.

Acknowledgements

This research was done with the support of the Universidad Michoacana de San Nicolás de Hidalgo and a grant from the Mexican Consejo Nacional de Ciencia y Tecnología.

Chapter 14

To be for (an)*other*

The Caribe Hilton or ambivalence as presence in a United States colony

Luz Marie Rodríguez

> [. . .] every reading and interpretation of a canonical works re-animates it in the present, furnishes an occasion for rereading, allow the modern and the new to be situated in a broad historical field whose usefulness is that it shows us history as an agonistic process still being made, rather than finished and settled once and for all.
>
> (Edward W. Said 2004)

Introduction: the idea of architecture in a 'non-nation' or the inner workings of ambivalence

In 1898, as an end to the Spanish–American War, Spain ceded its remaining colonies to the United States. Early in the colonial transfer and partly due to its Spanish heritage, Puerto Rico was appointed the role of the bridge between the Americas. By the mid-twentieth century that recurring and much exploited theme became a crucial geopolitical tool supported by the power of the architectural image. Modern architecture lent itself visibility and helped to promote Puerto Rico, first as a laboratory of democracy and then as its showcase (Tugwell 1947: 10; *Time Magazine* 1958). On the one hand, the Island needed to cater to potential North American investors to drive its development. But also within the context of the Cold War, the United States sought to consolidate hemispheric harmony by exposing the benefits that a political and economic partnership could provide to underdeveloped territories. Thus, during the 1950s, through its apparently thriving colony, the United States intended to discourage Latin American countries from alliances with totalitarian regimes by showing where democratic capitalism could lead.

The Government had adopted architectural modernism as a paradigm in 1944 through the Committee on Design of Public Works. Within that agency's agenda of designing for progress, this new architecture developed by the Committee intended, as Enrique Vivoni mentions, to establish an 'image of international sameness' (1995: 7).[1] While in many Latin American countries a modern discourse and aesthetic in architecture supported strong political and cultural national constructs, from the colonial standpoint of Puerto Rico, nationalism seemed a problematic concept. The investors the Island eagerly sought required guaranteed stable political environments (Moscoso 1954: 12). However, because the local political majority claimed that, for economic reasons, neither independence nor statehood was suitable at the time (Muñoz Marín 1992: 260, 267), the favoured tactic of the Popular Democratic Party was to downplay political nationalisms. So, by the 1950s, the advancement of a cultural nationalism was politically driven and sponsored in order, among other things, to counterbalance the prospect of assimilation that a permanent or prolonged association with the United States pointed towards. Despite the cultural handicaps that such a panorama supposed, local policymakers endorsed in 1952 the Commonwealth of Puerto Rico as the political liaison with the United States that would drive forward the crucial socio-economic development of the Island.

In the *Godkin Lectures* held at Harvard University in April 1959, Luis Muñoz Marín – the first elected Puerto Rican Governor – rationalized an implicit contradiction. For Muñoz Marín, the obvious obsolescence of nationalism legitimated a somewhat ambiguous definition of Puerto Ricans. The Governor outlined their political character as anti-colonialist, but still faithful non-nationalist citizens of the United States (Muñoz Marín 1959 in Reina 2005: 137). Let me underscore at this point that at the time the most recent political status of Puerto Rico had been, from 1898 to 1952, that of a United States' colony. Subsequently, 1952 onward, the Island engaged in the dynamics of 'coloniality', as defined by Walter Mignolo (2007: 32–33),[2] through the Commonwealth agreement. To this day, Puerto Rico remains an unincorporated territory of the United States via the Estado Libre Asociado (Commonwealth of Puerto Rico), a rather ambiguous terminology roughly translated as a free-associated state. Arguably then, government sponsored modern architecture was not openly nationalistic, as its objective appears to have never been the representation of Puerto Rico as a nation state.

The historiography of modern architecture in Puerto Rico lacks evaluation supported by postcolonial theories and/or geopolitical interpretations of its iconic works. The absence of that analytical framework prompts specific problematizations of the subject. This work considers the Caribe Hilton hotel as a pivot point in the tactics of cultural and identity negotiation between Puerto Rico and the United States. Reading the project as a vessel for a specific kind of power discourse and hegemonic interplay, the analysis positions it as a discursive construct framed by the convulsive tensions of difference that usually surround systems of domination. Within the scope of 'colonial mimicry' set forth by Homi Bhabha (2010: 121–131) the argument follows that in the active context of ambivalence that renders set mechanism possible (Bhabha 2010: 122–123), the Caribe Hilton seems to have performed as a shape-shifter of sorts between *sameness* and *otherness*.

But, why are twentieth-century references relevant for Puerto Rico's global presence in the twenty-first century? In a speech given in January 2011 at the University of Valladolid in Spain, Governor Luis Fortuño (2011) described Puerto Rico as historically a culture of integration. Fortuño suggested that the Island's multicultural heritage placed it at a vantage point within the notion of globalization. So, by identifying Puerto Rico as the convergent space of two of the main occidental civilizations in history – Spain and the United States – the Governor underscored the absolute relevance and possible benefits for Puerto Rico in the reclamation of its position as 'bridge'. Implicit in the Governor's statements were, of course, the advantages of a shifting identity within the contemporary global view of economics, politics, and culture.

On January 21, a day before Fortuño's speaking engagement, Madrid's newspaper *El País* published an interview where the Governor directly addressed the issue of Puerto Rican identity. After an inquiry about the cultural disadvantages of the Island becoming a state, journalist Verónica Calderón (2011) asked rather confrontationally if Puerto Ricans were Latin Americans or if characterwise they mostly align with the United States. Fortuño's reaction stressed in no uncertain terms that Puerto Ricans were Boricuas (the colloquial term derived from the native Taíno word Puerto Ricans often use as self-reference), Hispanics, and Americans (alluding to the United States, as is the norm in the country). He then followed the statement with a rhetorical question: 'Who do you love most, your mother or your father?'.[3] Clearly, the Governor tackled the problematic subject by advancing a foundational myth where, in his view, things were gained instead of lost. With the romantic notion of Spain as the mother country and the United States as the father nation, he intended to dismiss arguments that held statehood as the surest way to defeat Puerto Rican identity. What Fortuño did not consider at the time was that since both 'parent figures' are in fact, icons of colonial domination, his rather nostalgic approximation permanently ties Puerto Rico to representations of the subaltern.

Arguably, the reaffirmation of Puerto Rico's political, economic, and by now, cultural link to the United States, while managing a discourse that claims strong ties to a Spanish heritage, actually emphasizes the idea of an ambivalent identity as formulated by Homi Bhabha. Evidently, because of several political technicalities, the Island is no longer considered a colony. Still, Fortuño's profiling ratified the conscious and/or unconscious act of highlighting certain qualities that, despite being exhibited for self-gain, validated, nonetheless, an 'incomplete' and/or 'virtual' presence (Bhabha 2010: 123). That is, Fortuño confirmed, as perhaps inadvertently, the Puerto Ricans still rely on ambivalence to convey some 'image of identity' (Bhabha 2010: 64). That liminality, problematic as it may be in the particular contexts of Puerto Rican politics and cultural criticism, became fixed by a history of asymmetrical negotiations. Subsequently, specific ambivalent actions, such as 'to be *for* an Other' (Bhabha 2010: 64), were – and still seem to be – part of a certain survivalist attitude. Thus, if the Governor's speech and interview are any indication, it can be sustained that in the twenty-first century Puerto Ricans keep performing within the transformational virtues of ambivalence. I propose, then, that the project for the Caribe Hilton hotel (1946–1949) – the first in the Hilton International chain of hotels – might be helpful to understand how architecture operates within and assists in configuring such processes.

Readings on the competition for a 300-room hotel

What eventually became the Puerto Rico Industrial Development Company (PRIDCO) was created in 1942 to promote the Island as a possible business venture for the United States' investors. By dubbing it 'an industry without factories' (Compañía de Fomento Industrial de Puerto Rico 1949: 26), PRIDCO's director Teodoro Moscoso thought the logical complement to industrialization was tourism. But even with Moscoso's endorsement, the development of tourism did not figure in the programmatic priorities of the Popular Democratic Party (Maldonado 1997: 119–121). In spite of this, convinced that a hotel was crucial within the industrialization project, and despite the party's opposition, Moscoso advanced it as a business tool. He then persuaded his objectors by presenting the hotel as a place where PRIDCO would be able to suitably accommodate the prospective investors the Government invited to weigh the Island's business opportunities (Maldonado 1997: 121).

By 1946 Moscoso had already captured the interest of Conrad Hilton. For Hilton, the prospect appeared a highly appealing opportunity as the Government of Puerto Rico would assume almost the total cost of construction for the hotel that was to properly start his company's already projected expansion into foreign markets (Maldonado 1997: 121).[4] As the next step, an architectural competition was set in motion. Paradoxically though, the fact that the first Hilton International hotel was to stand in a United States' territory shows how colonial discourse as settled on difference was unavoidably implicit on the project. This 'international' Hilton was really a hotel in an-*other* place.

In a 1947 report on the viability of tourism, consultant Frank Martocci argued that because the Island did not ascribe to European or North American ludic preferences, Puerto Rico was not a suitable destination. So, in order to guarantee a market, Martocci advised hotels in Puerto Rico to cater to luxury and quality, while maintaining affordability. As other neighbouring islands had already established tourism as part of their economies, to him it seemed evident that reaching a desirable level of competitiveness would take a considerable amount of time. Hence, emphasizing as the only favourable characteristics for the industry those directly connected to the Island's topography and environment, Martocci also suggested taking advantage of the natural resources of the region. His solution: combine the positive aspects of the climate with a sea front development in the capital city, San Juan (Martocci 1947).

While on an inspection visit Hilton executives chose the site for the hotel (Maldonado 1997: 121–122). Overlooking Condado, one of the luxury residential areas in the country, the project would occupy a spot near the Spanish colonial San Geronimo Fort at the eastern point of the San Juan peninsula. For the competition Hilton contacted two Florida firms: Frederick Seelman of

Palm Beach and Robert Swartburg of Miami Beach. PRIDCO, on the other hand, invited the local architectural firms of: Schimmelpfenning, Ruiz, and González; The Office of Henry Klumb; and Toro, Ferrer and Torregrosa.

After the review of the surviving drawings,[5] similar floor plans in the Beaux Arts tradition seem implicit in the presentation perspectives from both Florida firms. In addition, Seelman and Swartburg preferred an eclectic style influenced by the Spanish Revival. Perhaps their intention was to distil a certain exotic image from the ever-fashionable world of Hollywood glamour (Figure 14.1) – an image which, as observed by architect Jerry Torres Santiago (2000: 130), appears to verify how almost in mid-twentieth century a collage of Hispanist and Orientalist decorative elements in a 'fantasy' setting were deemed 'the exotic'.

Compared to the entries by the firms working in Puerto Rico, the projects by Seelman and Swartburg clearly placed Puerto Rico as the ex-centric. Both entries seem discursive constructs within the mainstream stereotypical mindset of most metropolitan architects. These imports were no more than vitiated images based on an explicit definition of the Tropics as *other*. So, drawing from Edward Said's (1979: 12) rationalization of Orientalism, Tropicalism

e 14.1 300 room
, Florida entries
).

appears as a geopolitical fabrication intended, by way of a certain imposed imaginary, to (re)create, (re)define, and manipulate the difference the Tropics – or the tropical – supposedly conveyed within an asymmetrical socio-political structure.

From that perspective it is not unfounded to make a case for Tropicalism as a colonial mechanism where architects seem to have been key players. Thus, by reading Seelman and Swartsburg's projects as Tropicalist constructs, they reveal themselves as referent products. Through any one of them, Puerto Rico would have been easily recognized as the exuberant, colourful, irrational, libertine space of excess by comparison to the metropolis. Evidently these entries did not speak of the true Tropics – much less about Puerto Rican-ness – but actually validated the representation of Puerto Rico as subaltern.

For the 300-room hotel competition Henry Klumb[6] was the only architect to name his project. With the *Hilton's Tropical* he appears to explore and refine the idea of the Tropics as a resource as opposed to a problem.[7] Typically, architecture in the region had been approached as a combat with climate, but Klumb uses the Tropics as a design concept in order to create a specific kind of modern habitat. By including covered walks and a tropical garden as architectural features that also were to take advantage of the views and the breeze, the architect was advancing an-*other* Hilton. As the perspective shows, the architect reinforced the setting by a treatment of the façade that was to include tropical vegetation in plant boxes and vertical elements where crawling plants could thrive. Also, implicit in that drawing is the idea to harness the reflective quality of glass and allow the building to function as a mirror where the landscape would find its echo (Figure 14.2).[8] This was a hotel where difference, reworked as what historiography of modern architecture has termed Regionalism, would be explicitly emphasized without reading as a parody. That is, where *otherness* became a conceptual resource to design.

In contrast to the Florida entries, Klumb's proposal showed a modern aesthetic and a functional approach to design. For instance, the typical guest room floors showed the service circulation – central and enclosed – separated from that of the guests, which were peripheral and open. Thus, hotel guests would access their quarters from an exterior gallery into a featured space labelled in the drawings 'sitting and view room'. With that concept, Klumb was able to discard the balcony by elevating the floor level of the rooms from that of the circulation galleries. Then, by the introduction of a curtain wall, views were favoured while allowing for privacy. But the project was not completely innocent as from the colonial standpoint that solution, even if dramatic, seems to enable the colonizer's desire for separation. The window might also read as physical and psychological protection against the 'exotic'.[9]

The point is validated by the work of another architect. As Le Corbusier started to familiarize himself with the Latin American Tropic through his visits to Rio de Janeiro, in texts such as *Precisions* (first published in 1930) he began to discuss his ideas on the ways in which architecture could work within what he thought of as an unruly environment (Le Corbusier 1991: 244–245).

ure 14.2 Henry
mb, *Hilton's*
opical (1946).

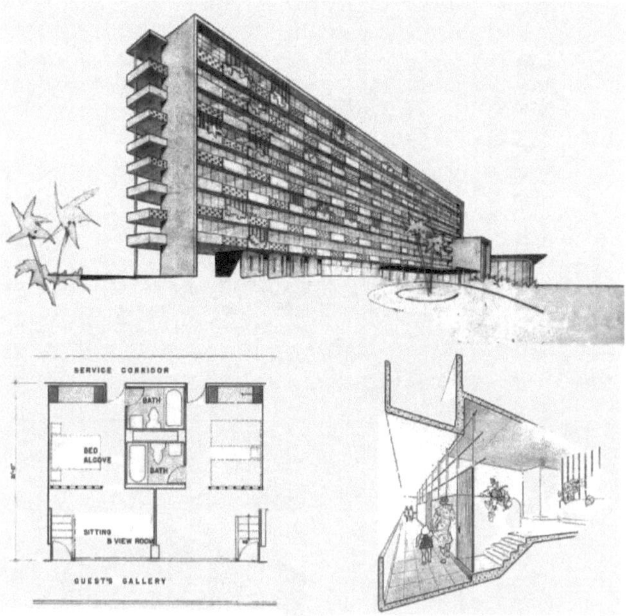

Yet in 1946, in the fourth volume of his *Œevre Complète* – a book that was not in Klumb's library – the architect explained how nature was to be allowed into modern buildings in tropical settings by using the windows as frames to the views (Le Corbusier 1995: 80–81). Le Corbusier then illustrated the statement with a line drawing of an interior perspective where, in front of a window – working as though it was a landscape painting – he placed the figure of a man, sitting while contemplating nature. For Le Corbusier that solution, also a 'sitting and view room', was the way to dominate the supposedly menacing nature of the Tropics.

The Caribe Hilton or ambivalence as presence: the complicated shifts of cultural negotiations

The Caribe Hilton, built by the Puerto Rican firm of Toro, Ferrer, and Torregrosa, was predominantly a horizontal rectangular volume with eight of the ten stories of the hotel for guest quarters. Contrary to Klumb's submission, the architects preferred individual balconies for each of the rooms. These are triangular cantilevers formed by the rotation – approximately 65 degrees – of the rooms' exterior glass partitions. A novel element in hotel architecture at the time, that solution allowed for ocean views from every room. Then, not only did a synthesizing element in the form of a grid inject visual impact to the façade while unifying the balconies, but also the reticular *brise-soleil* ensured protection against solar rays as it directed the breeze (Figure 14.3).

Figure 14.3 Toro, Ferre and Torregrosa, *Caribe Hilton* (1947).

Unquestionably, as the most published building of the Island at the time, the Caribe Hilton's international presence intended to exploit the aesthetic of the *new* while promoting a specific Puerto Rico: one supposedly modern, comfortable, and hygienic. Historian Ricardo Salvatore describes the disparity between North America and Latin America in practical terms by noticing implicit inequalities cemented on the idea of technology and consumer culture. As he indicates, for United States' travelers, comfort was a concept directly associated with certain services and products such as good hotels (Salvatore 2005: 283). Thus, what were considered deficiencies in the experience of comfort were in turn equated to underdevelopment. With that, the reasons why Moscoso was emphatic on his preference for a hotel that architecturally manifested an efficient modernity are aptly explained within PRIDCO's goals to introduce tourism as an industry while also seducing potential investors.

Conversely, Andrea Giunta (2005: 200–201) argues that in the post-World War II context, internationalism referred to the expansion of the United States' democratic-capitalist system. She explains how abstract art was chosen by museum elites as the technique best suited for its representation. By the same token, a similar discourse seems to have operated within the version of modern architecture codified as the International Style. As abstract art and 'abstract architecture' appeared not to be place specific, images that were not abruptly local or where such references would peak through as a secondary reading – that is, where internationality seemingly occupied a superior hierarchy – facilitated the diffusion of the imperialist-democratic-capitalist system because the discourse, although visually translated, was kept somewhat nonrepresentational.

For PRIDCO, the Caribe Hilton had to provide an image resounding of the North American qualities of Puerto Rico. Those were modernity and efficiency, as opposed to Hispanic traits such as an old-fashioned romantic charm, which was deemed unfitting at that particular instance (Moscoso cited in Ross 1966: 103–104). This Caribbean Hilton – notice that it was never the Puerto Rican Hilton – relied on the accepted metaphor of an unarguable progress by exposing North American-ness while, in a way, obscuring what was 'Hispanic'. Thus, by an architectural gesture, Puerto Rico seemed intended to 'equate' itself to the United States. An article published in the March 1950 issue of the *Architectural Forum* (1950: 102) attested to the observation as it remarked how the building lacked in local character. Also, in *Latin American Architecture since 1945* Henry Russell Hitchcock (1955: 57) seemed convinced that, even if tailored to the Tropics, the Caribe Hilton performed as a version of North American architecture.

Among the few references to Regionalism as Puerto Rican-ness within the building, stood the interior design of the guest rooms. As architect Miguel Ferrer confirmed in one of his last interviews, the consulting firm of Charles Warner and Harold Leeds recommended the use of elements and materials reminiscent of the Island's architectural traditions (Marqués Mera 1998). However, as Annabel Wharton (2001: 4) explains, in an effort to domesticate the exotic and show the guests an edited experience of *otherness*, the Hilton International hotels – for which the Caribe Hilton arguably was the prototype – reserved local expressions for the most intimate parts of the building. It follows that in the Caribe Hilton most of what could be defined as Puerto Rican was placed within the most private spaces – the hidden spaces, if you will – while the international and the tropical – non-interchangeable with Hispanic or Puerto Rican – were readily visible. Architecturally, internationality translated into the use of Le Corbusier's points – pilotis, free plan, free façade, roof terrace, and also the *brise-soleil*.[10] The Tropics were exalted by a real, not merely visual, open fluidity between exterior and interior spaces, through the landscape designs of Hunter Randolph and the articulation of the building as a multi-sensory / spatial experience as opposed to a rational / linear one.[11] In other words, internationality you saw, while tropicality you had to spatially experience with the involvement of all the senses in order to fully comprehend.

The social chronicle for the hotel's inauguration serves to measure ambivalence as a play on difference. In December 1949, Carmen Reyes Padró (1949: 1, 16) wrote in the local newspaper *El Mundo* how a metropolitan ambiance, reminiscent of New York or Hollywood, seemed to flourish inside the Caribe Hilton. Meanwhile, outside the building, the whole Eden-like quality of the landscape highlighted for the visitors the fact that they were in an-*other* place.

Arguably, the Caribe Hilton was that single space in the colony which aesthetically and ideologically established the site of modernization, internalization, and equalization in the same terms that Jorge Francisco Liernur (2002: 23) advances his definition for the metropolis.

As Bhabha (1994: 64) maintains, in the ambivalent definition of *difference, sameness* gets equated to being 'different from those who are different'. So, inscribed in the strategy that placed Puerto Rico as the bridge between the Americas, ambivalence played a pivotal role since as connector the Island was often described as resembling both Latin America and North America, while being unlike either.[12] Consequently, in the Puerto Rican case enunciated by the Caribe Hilton, to explicitly deemphasize the Hispanic while highlighting the supposed United States' character of the colony seemed intended to turn difference into sameness.[13] The building situates, then, a distortion based on what Bhabha (1994: 122–127) terms 'colonial mimicry' as it displayed the simulacrum of *sameness* first. But shape-shifting also came into play through the discourse of tropicality as colonial negotiations sometimes needed to underscore references to *otherness*.

Final words: identification

For Conrad Hilton, the modern American aesthetic emphasized by the international hotels eventually became an essential instrument in a personal political project. Framed within the Cold War, in Hilton's view each hotel was a contained representation of the United States able to show the benefits of democratic capitalism (Hilton 1957: 390, 330, 327). For PRIDCO, the hotel was a necessary tool since it would give the first impression of Puerto Rico. Paradoxically, as has been shown, the Caribe Hilton did not refer to the architectural articulation of the Island's identity but rather it set the stage for the specific system of identification.

As assigned by Bhabha (1994; 64), identification works as an impression of identity or an artificial representation produced '*for* an Other'. As a fabricated version of identity, in the Caribe Hilton, identification mostly meant to 'appear' the same but still stand in *Otherland*. In that context, ambiguity assisted by 'ensuring' the potential investors and tourists the Island's 'modernity' by catering to their specific interests. Those were, as Martocci consigned (1947), to stay with their peers in a lavish, semi-enclosed, yet exotic setting. In addition, as in the 1950s the Island was the model of modernization for Latin America within the canon of democratic capitalism, the international aesthetic served as the infallible image of progress – but only if aligned to the United States' political and economic ideologies.

At a time when the political relationship with the United States did not allow for the explicit expression of 'Puerto Rican-ness', the idea of the Tropic operated as a working concept for *otherness*. That way, the schematics of domination were kept in place as relations with Latin American countries – which favoured United States' political and/or economic prospects – exempt from nationalistic/

emotional ties could be encouraged. The Caribe Hilton was the architectural platform where Puerto Rico began staging a visible and much publicized transformation. As a performative construct identity representations within the building shifted from *sameness* to *otherness*. So arguably the Caribe Hilton was never the architectural representation of Puerto Rican identity but its fictional reference in a context of non-confrontational colonial negotiations.

Chapter 15

The mirror of territorial identity in Singapore professional architecture 1923–1969

Colonialism, nationalism, separation and independence

Raymond Quek

In a globalized world, nominal reference and titles differentiate and contribute to spatio-temporal identity formation. Changes therein therefore challenge the stability of territorial identity. Singapore and Malaysia are two nations that share not only much common history, but also many territorial claims, physical and otherwise, that have had parallel and often competing development as the pivot of colonial British Malaya prior to their independence. These parallel similarities and cultural dualities stake and outline claims of identity within broad phases of political paradigms over a span of time: colonialism, nationalism, separation and independence; and illuminate the question of territorial identity.[1] Vacillating conditional phases within the Singapore professional architectural society, its iterations and journals, and the correlation to changing guises in contested territories of identity offer insights to the nature of challenged identities in a precarious global world that continues to fragment or reconfigure.[2]

Preparations for an Architects' Ordinance in British Malaya started in earnest in April 1917 through the efforts of Campbell Oman[3] and a few others to regulate building construction practices and to contribute to the building of a city as a concerned professional collective. What Oman and his team perhaps did not realize is that they were outlining a stake of territorial claim in the colony. In November 1920 the municipal commissioners were given powers to classify qualified persons as 'architects'; this was not however adopted by the municipality of Penang, and hence did not have full application over all of British Malaya (JSSA Mar 1924: 34).

Singapore Society of Architects

On 22 March 1923 the Singapore Society of Architects (hereinafter SSA) was formalized with the main concern of professional architectural registration; around the same time similar concerns were being raised in London, and elsewhere in the British Commonwealth. The membership was mixed, a third of them being local draughtsmen who operated as 'architects'.[4] The SSA wrote to the RIBA on 18 April 1923, and their application for alliance was approved on 25 June 1923 (JSSA Dec 1923: 38).

The SSA published a monthly professional journal detailing accounts of negotiations with the municipal government. The first issue of the JSSA went to press in December 1923, and sported the Society's crest on the cover, which depicted a sub-Saharan maned lion behind a singular Corinthian column, ringed with the Society's name (Figure 15.1a). This is clearly a play on the RIBA crest which has two lions and a column. Oscar Wilson, as Assistant Honorary General Secretary, was acknowledged as the editor of the JSSA (JSSA Mar 1924: 32–34). However, there were no editorials in the JSSA issues, and it was odd that in issues 2 and 4 successive SSA presidents took to corresponding to the society via its journal as if disconnected (JSSA Jan 1924: 32–33). The JSSA comprised republished articles and a few original contributions that were topical. There was nothing much to publish of local work because of levels of apathy, and a nagging concern about whom the SSA were actually representing.[5] It is clear that the effort toward professional registration made progress as the municipal government approved the final draft of the Bill in 1926, and the Architect's Ordinance came into effect on 1 January 1927. In 1929, the Singapore Society of Architects Incorporated (hereinafter SSAI) was formalized with a register of 46 licensed persons. Most of the original signatories to the SSAI were Fellows of the RIBA who followed similar developments in London very closely (JSSAI June 1929).[6]

Ho Kwong Yew[7] designed the crest of the SSAI, winning an open competition in 1928 (TMA Dec 1935). The design involves a mischievous replacement of the one of the two lions buttressing the column in the RIBA crest with a Malayan tiger, a modification also of the earlier SSA crest. The convenience

re 15.1a and SSA crest gner unknown, circa (15.1a). SSAI crest gned by Ho Kwong circa 1928–29)).

lent by Singapore's claim to the lion via the legend of the *Singha*[8] completes the design, which is clearly derivative and allied to the RIBA as desired, but concurrently depicts the feline symbols of two entities of British Malaya, both contesting for centripetal significance, seemingly in equipoise though the lion wears a crown whilst the tiger does not (Figure 15.1b).

When the Society of Architects was incorporated it was recorded that membership comprised almost all of the persons operating as 'architect' in the Malay peninsula. The SSAI saw itself as the pivot of the peninsula, presuming significance in the greater hinterland. It is a fact that there was a greater density of these professionals based on the island of Singapore, many of whom had branches and offices in Kuala Lumpur, Ipoh, Penang and Malacca. It is also true that there were practices solely based in Ipoh or Penang that had little to do with Singapore and the SSAI. The Straits Settlements era governance of the municipal regions was dispersed in Singapore, Penang, and Malacca, and the Malay States, as were the governance over licensure and control of building permits. A completely duplicate set of Government officials, based in Kuala Lumpur, operated under Federated Malay States (FMS) Ordinances (JSSAI Apr 1930: 2).[9] However, the professional society and the legal register of architects were quite separate entities. The Board of Architects managed and controlled jurisdiction for the municipal commissioners. The interaction of daily life between territories in the Malay states and the municipalities was common and normative, and therefore overlapping claims in these territories were taken for granted. The anxiety to have clarity of belonging, inclusion, participation and territorial claim was at the onset a dilemma; it was later noted that the initial name of the SSA was mooted to be *Institute of Architects of Malaya* when the society started in 1923 (JSSAI Feb 1931: 2). This early concern over the tail wagging the dog and the suggestion to adopt the IAM name were also echoed by the RIBA in correspondence to the SSA in 1924 for fuller representation of British Malaya as opposed to the limited municipality of Singapore (JSSA Mar 1925: 2–3). The SSAI set out immediately after its formation to organize branches in Kuala Lumpur and Penang, and to compete for work 'up-country' (JSSAI Feb–Apr 1930). There was no parallel professional society in the Malay States, where the regulation of building permits by the municipal or territorial governments seemed to suffice. A branch of the SSAI was explored and eventually organized in Kuala Lumpur in 1930 (JSSAI Feb 1930: 2–3; PETA 1955: 35).[10]

Singapore Society of Architects Incorporated and the Institute of Architects of Malaya

In the view of the first Honorary Secretary of the SSAI Oscar Wilson, the critical territory between everyday and collective universal identity was the condition of modernity, that which yoked the vernacular and the universal and allowed growth. He singularly undertook the responsibility to produce and edit the Singapore Society's new journal starting afresh as the JSSAI (JSSAI August 1930: 1).[11] He was editor before with the JSSA, but the new task of the SSAI was to forge a city with a modern architectural outlook (Figure 15.2). The JSSAI was strongly independent in tone and the editorials revealed the drive of Wilson's modernizing crusade on all matters of practice, business and law but also revealed a benevolent sense of technology and information transfer in architectural design, practice bureaucracy and town planning vis-à-vis the state of the art of modern architecture. All this was pursued with a zealous yet aggressive approach as shown by a paragraph of his first editorial foreword:

> It is hoped that in time to come the shophouse fetish may be destroyed, and that building owners, and even building authorities may cease to think of building schemes only in terms of shophouses, and Singapore may cease to be a city of mean streets, and with a few notable exceptions, mean buildings; and much may be done by a Journal such as this in assisting the cause of town planning, in the development of vulgar monstrosities, which disfigure the suburban districts.
>
> (JSSAI Jun 1929: 1)

THE HON. SECRETARY.

e 15.2 Caricature car Wilson, drawn nis Santry circa

The JSSAI and JIAM were published with the address of Wilson's office, whereas the JSSA before them had no recorded address of publication. This sense of personal commitment and lack of neutrality is worth remarking, as it reflects the colonial expatriate's stake in the colony.[12] From the launch of the journal, Wilson's chief editorial targets were the slums in the city, the standard of health and the quality of architectural practice. Perhaps the most successful of his attacks was on the state of the urban slums amongst the shophouses. This was orchestrated in sequence through Wilson's editorial rants, George Duckworth's republished RIBA lecture 'Making, prevention and unmaking of a slum' and a series of original articles written with a *nom de plume* of *Townplanner*, clearly a municipal official disguising professional conflict of interest (JSSAI Sep/Oct/Nov 1929, Jan 1930).[13] A total of four successive articles by the *Townplanner*, more rants by Wilson against the municipal bureaucratic machine, an anonymous article on clearance of slums and 'The Slums' by John Angus (JSSAI Feb 1930: 9) were also published between late 1929 and early 1930 in a sequential series of monthly issues. The articles had the desired effect on the Singapore Improvement Trust to act sooner, and the first clearance of slums commenced in 1931.

The age of municipal governance in Singapore sits undecidedly between the extremes of a totally modernized mind where anything colonial is an anathema, and a pure colonial mentality that aimed to replicate British life and standards, remaining dutifully subordinate to the Empire. The oscillation of identity is also apparent in the typical expatriate architect of the time. This anxiety of belonging and the challenges raised by Wilson certainly made it a very modern encounter. The period of the 1920s to early 1930s was an era rich with modern challenges in architecture. Wilson did not acknowledge Hitchcock and Johnson's recognition of international architecture directly but published various trans-national buildings and ideas, and through the journal instigated a conversation about modern multicultural architecture in Singapore. In January 1933 Oscar Wilson and others were nominated unopposed as the IAM leadership (JIAM Jan 1933: 6) but in February 1933 an entirely different leadership, none of whom were nominated in January 1933, were instead elected unopposed (JIAM Feb 1933: 4).[14] Wilson went into retirement shortly after, and the journals without his editorship were issued with little original comment, observation or contention.

The Malayan Architect

This oscillation of belonging amongst the municipal colonial architects anticipates the nationalism to come after WWII. This ambivalence is seen also in the vacillating title of the professional society itself. We have noted prior recorded ambivalence in the SSA name in 1923–1924; it was not until 1931 that all agreed that the name should be Institute of Architects of Malaya (hereinafter IAM) to reflect inclusive relations with the hinterland, and the *Journal of the Institute of Architects Malaya* (hereinafter JIAM) served as the Institute's monthly print communication (JSSAI Feb 1931: 3). In July 1933 the Institute's journal was renamed again and became *The Malayan Architect* (hereinafter TMA). These changes reveal an entrenched unease with the condition of territorial claim to identity. From the Singapore perspective, inclusion vacillated not only as an idea, but also in the manner of its expression. The three TMA issues following Wilson's retirement as the institute's secretary contain no editorial comment nor original writing, only a handful of republished technical papers which appeared to have come from the reserve collection of Wilson's editorship (JIAM Mar/Apr/May 1933). By the fourth issue after Wilson's tenure as editor, the TMA had become desperate and issued monthly full-page pleas for original material and support of the journal (TMA Jun 1934–Oct 1934). There was no longer any concern for planning the city, removal of slums, or public housing. This journal had reverted to an almost pure commercial or technical press; there were no rallying editorials save for non-polemic announcements. In all there was no leadership that Oscar Wilson provided. Henry Arbenz, as president, introduced a journal committee and business manager to handle all the business editorial work for the TMA that was once handled by Wilson alone.

Whilst Wilson was concerned with the process of professional architecture, under Arbenz's leadership as president, this concern was largely eradicated for the consequence of architecture: the completed buildings themselves. In publishing the consequence, any possibility for learning or re-orientation during the process was no longer there, and Wilson's agency of agitation on behalf of the collective was lost. The January 1935 issue was completely dedicated to publishing the newly completed RIBA headquarters at Portland Place, London, a rather convenient subject for the poverty of content, but an editorial decision that also betrayed imperial sycophancy. Other projects published in subsequent issues also emanated from the temperate British Isles, placing the referential centre clearly in the heart of Empire. The contrasting disparity between immediate proximity and remote distance in terms of relevance betrays the missing middle ground of genuine collective concern. Letters from the changing RIBA presidents clearly revealed an unquestioning acquiescence to a subordinate role. One letter in particular, by Percy Thomas, was published to ask all members of the IAM to contribute a minimum of five shillings towards the costs not only of the portrait of Sir Ian McAlister, but also a copy for his wife (TMA Sep 1935).

In the post-Wilson phase, the journal had become non-confrontational, and government architects such as Frank Dorrington Ward and other individuals, some alienated by Wilson, started to surface and feature in the IAM and its journal. Whilst Wilson typified the colonial who staked all in the territory, the new QJIAM tone reflected an ambivalent neutrality. Lacking insight to garner more relevant or topical material to progress the profession, the new journal committee decided to feature a monthly section on Malayan personalities in architecture, commencing with Arbenz himself (TMA Apr 1935–Jan 1936). Ho Kwong Yew was the only native architect to be featured in the series that celebrated the colonial architectural contribution to British Malaya (TMA Dec 1935). He certainly lived the life of a gentleman architect who gave all to his territorial identity,[15] he indulged in photography and was wealthy enough to be immortalized in watercolour sporting a necktie in pre-air-conditioned times with wife and young son, and a hand on architectural plans (Figure 15.3). Henry Arbenz was of Swiss origin and was not a Fellow of the RIBA, but he was welcomed into the fraternity of Anglo-Singapore architects since 1924. He was an architect but also the Honorary Consul of the Swiss for British Malaya (TMA Apr 1935). Like Ho, Arbenz was an outsider who was accepted into this ambivalent band of displaced colonials.

With the commencement of the war, the TMA ceased publication in November 1941. Despite its professed mindset to be inclusive of the hinterland, it was for all intents and purposes focussed on Singapore as is evidenced by the sheer scarcity of content related to matters north of the municipality. Singapore was the perceived centre of the IAM of British Malaya: the crowned lion against the uncrowned tiger in the crest stayed through all issues of the JSSAI, JIAM and TMA. It was not 'corrected' until the postwar era (Figure 15.4).

Postwar societies and journals

When the war ended the society and the journal recovered in a new world of mushrooming independence: the formation of Israel, the independence of India and Pakistan no doubt spurred independence in the region (QJIAM Mar 1951: 13; PETA 1955: foreword). The critical distance between everyday life and universal life, which for Wilson in 1929 was reduced to the point of friction, became the case for general populace. With the end of the war, questions of state and nationhood, allegiances to socialism or communism, were choices thrust upon

Figure 15.3 Portrait of Mr and Mrs Ho Kwong Yew and Ho Kok Hoe, 1939, painted by Xu Beihong. Collection of Ho Kok Hoe.

ure 15.4 Governor John Nicoll, Patron IAM, and IAM crest, dified design of SSAI st, 1931, with claims ineage to SSA in 3.

H. E. THE GOVERNOR OF SINGAPORE, SIR JOHN NICOLL, K.C.M.G. Patron of the Institute of Architects of Malaya.

THE SEAL OF THE
INSTITUTE OF ARCHITECTS
OF MALAYA

The President (1952-1953) Mr. J. M. Fraser, A.R.I.B.A., M.T.P.I., was the first President of the Institute to wear the New President's Badge of Office and Collar, when presiding at the Annual Dinner of the Institute on 6th February, 1953.

the polity. The Federated States of Malaya were re-organized in 1948 to include to former municipalities of Malacca and Penang, giving the hinterland clearer stakes in territorial identity. In 1949, the Federation of Malaya Society of Architects (FMSA) came into being in Kuala Lumpur. Though the FMSA had its origins as one of the Singapore based IAM's satellite northern branches, the Institute of Architects of Malaya in Singapore operated separately. The IAM's days as the tail wagging the dog were numbered. Singapore remained a crown colony, with a bubbling independence movement that emerged alongside a triumphant sense of colonial governance. The British maintained they won the war for the Empire and the colonies, whilst the Singaporean memory was clear that the British surrendered the island to a Japanese bicycle regiment and that the Americans won the war for the world with the nuclear bomb.

The de-colonization of British Malaya was inevitable. It might be noteworthy to remark that, as with several forms of organization in the region at the time, the divided and differing regional governance was reflected in much, if not the entire concomitant substructure of collective organization. Hence British Malaya was always the reference of greater collective ambition, whether an initiative commenced in the colony of Singapore, the settlements of Malacca or Penang, or elsewhere in the Federation; and more remarkably whether the initiative was inclusive or exclusive of the whole of Malaya.

By October 1951, an ordinance had been passed to allow architects registered under the Ordinance of 1941 in the Crown Colony of Singapore, the Settlements of Malacca and Penang to be registered in the Federation (Architect's Ordinance 1951). The journal also listed the members of the Institute, according to their professional status, student status or geographical disposition with the now quaint term 'up-country'. The relationship with the hinterland so defined, the list included 'up-country' members mostly based in Kuala Lumpur, rather than Penang or Ipoh. The *Quarterly Journal of the Institute of Architects of Malaya* (hereinafter QJIAM) commenced in March 1951.[16] The language and tone of the QJIAM editorials and structure were remarkably imperious, compared to the language of the 1930s, which seemed more engaged with the challenges. The editor was T. H. H. Hancock, who worked with the PWD.[17] We might also remark that though the readership was limited to those practicing immediately in the locale, the content reflected a strong colonial anthropological distance that revealed a lack of genuine immersion in the municipality. Unlike most colonials in the pre-War era, post-WWII era officials were largely newly arrived and sensed a short sojourn in the tense politics of the time. George Edgar Magnay, a member of the council and a future president of the Institute, and at the time a Planning Officer of the Singapore Improvement Trust, published a report on the exhibition 'A study of the Homes on Singapore'. His distance is shown in his remark of pure otherness: '. . . what at first appeared to be an ambitious experiment was justified by the interest shown by members of the public in the affairs of the town' (QJIAM 1951: 17–24). Whilst such an exhibition or study might be useful for the newly arrived expatriate, the public exposition in 1951 of such a basic survey was highly questionable and naive; Magnay showed the Singaporean public their very own homes without any particularly new insights. The postwar colonial arm gestured and postured more than it rolled up its sleeves and actively staked a claim in the region. However the QJIAM era introduced a positive interest in a non-temperate architecture, a worldwide phenomenon at the time.[18]

The gratuitous publication of photographs of various IAM presidents and the colonial governor is noteworthy of the distance between the ground and the top (Figure 15.4). In the history of the SSA, SSAI and the pre-war IAM, such gratuitous non-architectural photographs never appeared. In volume 3, the QJIAM begins acknowledgement of a patron in the form of the colonial governor in its contents page, a return to the practice of the colonial JSSA. The dedication of full pictorials to colonial leaders in regalia by this time can be interpreted as a purely theatrical claim of colonial control against a growing independence movement.

In June 1955 the FMSA finally launched PETA, a journal specifically for the Malayan Federation, with Stanley Merer as editor. Hancock, QJIAM's editor, was a member of the both IAM and FMSA councils, an ambivalent position afforded to the expatriate more easily than to a native of the region who had to face an increasing demand to declare allegiance towards one territory or the other. In 1955 significant changes in the region took place; native political parties had in effect taken over governance of the municipality and the Malay States. David Marshall and later Lim Yew Hock formed the first non-colonial elected Singapore Governments. The British colonial governing presence was effectively vacated in the political realm and so it was with the professional society. The QJIAM suffered: Volumes 2 and 3 were limited to two issues, and Volume 4 published four issues over two years, 1954 and 1955. The last 1955 editorial noted these changes of government and remarked that architecture in Singapore would have to become 'an architecture of economy' (QJIAM 4/1955: 3). No issues of the QJIAM were published in 1956 reflecting a tumultuous year in politics.[19] In 1957, Volume 5's two issues carried on with an additional prefix: 'Plan, the Quarterly Journal of the Institute of Architects of Malaya' (QJIAM 1 and 2/1957). Whilst the title of the QJIAM was tampered with, the editorial team did not seem to be disturbed by the promise to publish quarterly, which they never fulfilled other than in 1951. This lost sense of the temporal was perhaps an anticipation of a forlorn colonial project. It would seem to be highly remarkable that the JSSA, JSSAI, JIAM and TMA were published monthly.

The Federation of Malaya gained independence on 31 August 1957, and this did not include Singapore, which technically remained a Crown colony. Whilst from the very beginning of the SSAI, the need to establish inclusion of all that which once constituted British Malaya was a driving concern, the FMSA had no such qualms. There was great political uncertainty as there was indigenous communist insurgency to contend with, and negotiations for self-governance status stalled due to urban riots. No one really knew whether to look back to a safe colonial era or look forwards to an undefined future, and many were uncomfortably caught in the transition.

The ambiguity of perception is best seen in the case of Koh Cheng Yam. Koh was educated at the Architectural Association, London, and was by the 1950s a partner at Swan and McLaren. He was also the only native Singaporean to be president in a postwar IAM council, though he never enjoyed the privilege of published portraiture in regalia afforded to the colonial presidents before and after him. As president he harboured a sense of colonial equity, which he sought to defend. Victor Chew,[20] erstwhile employee of Swan and McLaren, recalled scathingly how Koh could absorb and accept racism, and bought into the postwar colonial masters' mentality. When Koh was made partner at Swan and McLaren, he was described to be the 'first and the last Chinaman to be a partner' by a colonial at the office, and was promoted only because he had offered care to his imprisoned colonial employers during the Japanese occupation (Chew undated). This is much more of a dilemma in reality for the conservative minded Koh, who tasted a sense of colonial equity, whatever others may have thought, and was certainly entitled to take his political position.

The IAM disintegrated and gave way to the new Society of the Malayan Architects (SMA) that shared the dream of the radical independence and nationalist movements in 1958. Seow Eu Jin, who was once invited to replace Koh as president of the IAM, chose to ignore that invitation and join the newly formed institute driven by younger Singapore-bred, as opposed to expatriate, architects. Neither Koh nor the colonial members of IAM show in SMA's first register and Koh thought little of the SMA (Chew undated). Victor Chew drafted the constitution and rules of this institute, the so-called Society of *Malayan* Architects of *Singapore* (hereinafter SMA, my italics). The members of the SMA collectively found distaste with not only second-class status in their own country, but also in the same subordination to the RIBA and the colonial structure of the IAM (Chew undated). The *Journal of the Society of Malayan Architects* (JSMA) was the 'official organ of the society' (JSMA 1 / 1958: 1). Ho Kok Hoe, son of Ho Kwong Yew, won a competition to design the crest of the SMA. The IAM published a final issue with a name change in October 1958, in the same year as the first issue of the JSMA (JTIAM 1 / 1958). The IAM had effectively run its course and finally dropped the promise to publish quarterly in its title; however, it is noted that its singular 1958 publication is Volume 6, Issue 1. Further issues must have been seen as possible, but never materialized, in the same way that the Crown colony might have possibly continued but did not.

The frequent names changes in the Singapore professional architectural journals reflect these very confusing phases of territorial identity and anxiety within the municipality in relation to hinterland and motherland; Empire, Federation, self-governing state and independent nation (Figure 15.5). Singapore announced the acquisition of self-governing status from Britain in 1959, but lobbied hard to be part of the Malayan Federation. As if to show inclination, Volume 2 (1959) of the JSMA adopted a linguistically local prefix: *Rumah*, the Malay word for *domus*. It was published as *Rumah: Journal of the Society of Malayan Architects* (RJSMA). This title lasted two issues until Volume 3 (1960). The political will toward identity was not very clear. Volume 3 featured articles such as 'The Malayan Touch' and Lim Chong Keat's riposte, titled 'The Malayan Architect on Trial' (RJSMA 1960) to the *Architectural Review's* (hereinafter AR) features on the region (*Architectural Review* Jun 1960). The authors of the AR articles, Lincoln Page, a former SIT architect; and Jules Poesner, who operated in 'the Federation', were architects returned to Britain and whose levels of understanding clearly riled Lim Chong Keat despite their years in the region.[21]

All the more curious was the inclusion of the high-profile national projects, the National Library and the National Theatre of Singapore. Ironically Singapore was not yet technically an independent nation state, and could in no way be understood as an ethnic singularity reflecting a nation of people; it did not even have self-governing status when the National Theatre and National Library were hatched as political initiatives in 1958. The foundation stone of the library was laid in 1957 and it was built by 1960. The theatre project was conceived in 1960, built with maddening urgency and operative by 1962, before the roof was completed (Quek 2012b). In the greater Peninsula, the Malayan government commissioned several national projects as well. Many of these were sporting facilities and most were completed at a much slower pace (Lai 2005). The projected purpose for the theatre was to foster nationalism and make one nation out of the ethnic groups. A sense of duality prevailed and no one sought to question why Singapore was pursuing national projects such as the theatre and library when it was not independent of the Crown, not yet part of the Malayan Federation nor did it have any unique ethnic or historical claim. A large portion of the populace was recently immigrant and had come from elsewhere in Asia. No one seemed to question the nominal confusion as the ties of kinship were spread across territories. Singapore was and had always been a territory

gure 15.5 Journal es, pre-war from 3–41.

managed by political apparatus from the days of the Straits Settlements, and has always carried a nebulous sense of territorial identity as we have shown.

RJSMA and RJSIA editorial records show that there was active concern about the demise of the IAM, and they actively sought to unify with the FMSA (RJSMA 1961; RJSIA 1966). Like the high-profile national building projects that were urgently completed, the SMA sought to unify with the FMSA repeatedly. In 1961, the society decided on another name change and quietly Volume 4 became *Rumah: Journal of The Singapore Institute of Architects* (hereinafter RJSIA), emphasising Singapore rather than Malaya. Britain released Singapore into independence on 31 August 1963, and by 16 September 1963 Singapore finally joined the Federation of Malaysia. The editorial of RJSIA in December 1963 nonchalantly noted:

> The recent development of the Society of Malayan Architects into the Singapore Institute of Architects represents a re-unification of professional groups in Singapore. The Institute of Architects of Malaya has dissolved itself to become part of the new Institute, which can now act as one body in the interests of the architectural profession.
>
> (RJSIA 1963: 2)

In 1964, Singapore was part of the Federation of Malaysia and *Rumah* was published as a volume of built work and projects speculated for Singapore, some brazenly celebrating Singapore's supposed nationalism. *Rumah* did not publish in 1965 but records claimed discussions were ongoing on the proposed unity of the now newly named SIA and the FMSA (RJSIA 1966: 3).

The political merger did not last and Singapore was ejected in August 1965, becoming fully independent. Commencing in June 1966, ten months following republican independence, a new bi-monthly journal, the *Singapore Institute of Architects Journal*[22] (SIAJ) was published by the Singapore Institute of Architects. *Rumah* Volume 8 was also published in June 1966, coinciding with the first SIAJ, with which it co-existed in an odd and haunting duplicity, a remnant of a bygone era of expired contested political will shadowing a new political will to survive sudden political independence. The FMSA became the *Pertubuhan Akitek Malaysia* or The Malaysian Institute of Architects in 1967. It must be mentioned that the entire clamour by the postwar IAM, SMA and SIA to include the territories of the north was in reality an approach unrequited, as the FMSA did not harbour such concerns and never reciprocated the idea of unification or merger in public, as much as the Malayan government did not actively seek to include Singapore into the Federation. It was Singapore that aggressively sought to be incorporated, though it preferred to understand the political inclusion it sought as 'merger'. It was the Singapore based institute and its journal that confusingly vacillated in nominal and territorial identity, whilst the FMSA remained clearly firm in its bearings.

After three years of absence, *Rumah* was last published by the Singapore Institute of Architects in 1969 as a single volume, and contained little commentary except for a short editorial note and a presidential message by Rex Koh who noted: 'This in a way depicts the genesis of Singapore's architectural achievement which is progressing parallel with the growth of our country' (RJSIA 1969)[23] (Figure 15.6). With that, what followed was a complete and literally silent pictorial volume of projects in Singapore completed by Singaporean architects – perhaps on matters of territorial identity there was nothing left to say.[24]

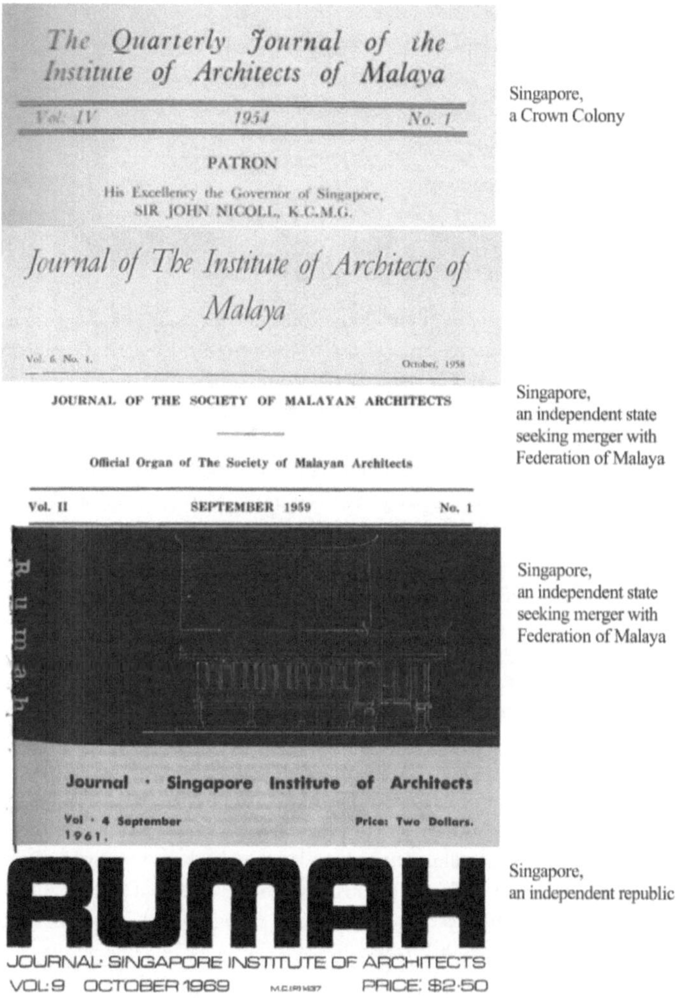

ure 15.6 Journal
es, post-war from
1–69.

Notes

Chapter 2 Buildings and the performance of identity

1. I use the gazetted place name, Bennelong Point, in this essay. Its indigenous name was recorded with various spellings at the time of the British arrival, is recorded as Bennelong's Point for many years and then becomes objectified as 'Bennelong' Point. The NSW Geographical Names Board in 2005 gave the point dual assignation with its geographical name recorded as Dubbagullee and its place name as Bennelong Point, see http://www.gnb.nsw.gov.au/name_search/extract?id=ujKqFxWA (accessed 13 January 2012). The working party involved in the reassignment process included members of the Aboriginal community and language experts and the process of arriving at original indigenous place names is recorded by Val Attenbrow, 'Aboriginal Placenames in Sydney Harbour, New South Wales, Australia: A Dual Naming Project' available from: http://www.ncl.ac.uk/unescolandscapes/files/ATTENBROWVal.pdf (accessed 13 January 2012). 'Bennelong' is the most common spelling given but his name is recorded differently in many sources. In the customary manner, Bennelong was known also by a number of other names – see, for instance: Clendinnen (2003, 103) Smith (2001: 159).
2. Keith Smith (2001: 51) and Inga Clendinnen (2003: 110), both of whom have written thoughtfully argued interpretations of the incident, construe it as a ritual spearing, organized by Bennelong. Both consider him to have been a keen strategist and suggest the incident was aimed at spilling British blood in a payback for wrongs the British had committed, and thus consolidating his position amongst his own people.
3. For instance, a map of the location sourced from Google Maps, http://maps.google.com.au/ (accessed 3 January 2012) identifies it as the Sydney Opera House.

Chapter 3 Emplacement, embodiment and ritual

1. These include: *shoji*, an inner translucent rice paper screen; *amado*, a middle solid wooden screen; and *sudare*, and/or an external timber lattice (Nitschke 1993: 85).
2. *Agate kudasai* is not a phrase which can be found in the dictionary, but does exist within common usage.
3. The word *traditional* is highlighted in italics throughout the text; questions regarding its use in architectural discourse will be considered later in this text.
4. The projection of a totalizing summarization is problematic. Obviously any culture is not homogenous, and includes permutations and contradictions. Various discourse as well as dialogue with Japanese living in Japan and the UK however evidences a commonality of shared ideas, practices and values regarding inside and outside.
5. *Shikii wo matagu* is not generally applied to public buildings and places; there is no inside, as the building is part of the wider community, i.e., the outside.
6. These include: *i(tsu)tekimasu*, as said by those departing; *i(xtu)tera(xtu)shai*, as said to those departing; *tadaima*, as said by those arriving; and *okaerinasai*, as said to those

arriving (Digital Daijisen 2011d, e, f and g). (The characters in brackets are not pronounced.)

7. Two aspects here warrant caution; the first is the use of the word 'Western', and by implication the term 'Eastern'. Such terminology has historically been utilized as a cultural point of self-reference to distinguish the 'other', that is, that which one is not. Any uncritical use of the terms overlooks both how they have been manipulated by both East and West to serve cultural, economic and political ideologies, and the intrinsic nature of culture not as fixed entities but rather as transient, shifting and fluctuating (Brown 2011; Brown and Maudlin 2012). The second note of caution is regarding slipping into a perspective of cultural relativism in attributing a greater presence of the body to non-Western philosophy (Noland 2009). Any negation of the relevance of the body in a Western context would fail to acknowledge a significant shift in the last 50 years on an understanding of the body as a site of cultural experience and meaning. (The above dates this shift back to Merleau-Ponty's seminal *Phenomenology of Perception*; Parkes suggests that Western thinkers' development of philosophies of the body originate in the 19th century with Schopenhauer and Nietzsche (Parkes 1995: 81)).
8. This discussion draws on Berque's discussion of Karatani Kojin, notably Kojin 1980.
9. A wariness of words like *tradition* or *traditional* is necessary. Equating *tradition* with 'posterity' or 'orthodoxy' reduces our understanding of *tradition* as a creative, adaptive and active process that is founded on and enables progress and change (Vellinga 2006). For example, the home that is typically celebrated as *traditional* Japanese was 'not an essential, unchanging entity, but one which came into existence over a long period of time' from the 1600s up through the mid 1800s (Cornell 1997: 29). In place of limiting and marginalizing definitions, we need to recognize tradition within a more inclusive understanding of place and identity (Brown and Maudlin 2012).

Chapter 5 Spatial memory and identity in the work of Istanbul/Vienna artist Canan Dagdelen

1. An earlier version of this essay was published at HTTP: <http://canandagdelen.com/wordpress/wp-content/uploads/2012/05/LLefairve_CD_on_Space_and_Memory_english.pdf> (accessed 11 March 2013).
2. See Noever, P., Witt-Dörring, C. and Dagdelen, C. (eds) (2004) *Canan Dagdelen: Yurt Tutmusch Dot*. Vienna: MAK, Austrian Museum of Applied Art, with contributions by Peter Noever, Christian Witt-Dörring, and Brigitte Borchhardt-Birbaumer, published on the occasion of the same-titled exhibition.
3. The idea of 'strangemaking' was coined by Victor Schklovsky in *Art as Technique*, first published in 1917. It is a device Alexander Tzonis and I have associated with critical regionalism in architecture.

Chapter 7 Architecture and memory

1. This exercise took place in Berlin in July 2012, and was realized together with Jonathan Wren, a second year student of Architecture at Nottingham Trent University, who was the recipient of the SPUR studentship award from Nottingham Trent University.
2. See Nora, Pierre (1996) *Realms of Memory: Rethinking the French Past* (New York: Columbia University Press); Halbwachs, Maurice (1980) *The Collective Memory* (New York: Harper & Row); Samuel, Raphael (1994) *Theatres of Memory* (London: Verso); Wilson, Keith (1994) *Forging the Collective Memory* (Oxford: Berghahn Books); Huyssen, Andreas

(2003) *Present Pasts: Urban Palimpsests and the Politics of Memory* (Stanford: Stanford University Press); Jordan, Jennifer A. (2006) *Structures of Memory: Understanding Urban Change in Berlin and Beyond* (Stanford: Stanford University Press); Rothberg, Michael (2009) *Multidirectional Memory: Remembering the Holocaust in the Age of Decolonization* (California: Stanford University Press); Landsberg, Alison (2004) *Prosthetic Memory: The Transformation of American Remembrance in the Age of Mass Culture* (New York: Columbia University Press); Macdonald, Sharon (2009) *Difficult Heritage: Negotiating the Nazi Past in Nuremberg and Beyond* (New York: Routledge); Logan, William and Reeves, Keir (2009) *Place of Pain and Shame: Dealing with Difficult Heritage* (London and New York: Routledge); to name but a few.

Chapter 8 Identities in the architecture of Mexico City

1. I would like to thank Nottingham Trent University for supporting my research and to Franca and Leonardo for their patience, encouragement and unwavering help.
2. There are various ways of understanding and defining globalization, which will depend on the particular dimension that is used as a keystone; globalization is a phenomenon formed by numerous aspects and relations, all defining the current global context. It is characterized by an economic neoliberal agenda driven by economic profit and movable capital, which has weakened the nation-state and challenged the traditional notions of national borders and identities. Globalization is also defined by the extension of social and cultural relations across the globe, partly aided by technological advances, but also by the movement of people between countries and other migrations. Individuals' mobility creates a sense of decontextualization, deterritorialization and cultural transaction. Broadly speaking, globalization should be conceptualized as a phenomenon that involves the expansion of political, social, cultural and economic relationships and occurrences beyond the national borders (El-Ojeili and Hayden 2006: 12–14).
3. The Constitution of Cadiz was promulgated on 19 March 1812 in Cadiz, Spain. It responded to the French invasion of Spain and had a significant impact in the viceroyalty of New Spain. It curtailed the political and social powers of the monarchy, accelerating its disappearance and strengthening the local governments by fortifying society's role in the governance of the country. Its liberal ideals were instrumental in the development of the independence movements in Latin America (Gamas Torruco 2012: X–XIII).
4. La Plaza de la Constitución has suffered various changes since the time of the Independence, however they have not impacted its basic urban morphology. The square hosted the equestrian sculpture of Carlos IV until 1822, this was replaced by other temporary uses, including the base *(zócalo)* from which it obtained its popular name. The Zócalo has accommodated commercial stalls, a kiosk, a tram ticket office and green areas.
5. According to the Revolutionary doctrine, the Revolution was the culmination of the historical process of emancipation, and construction of the nation that started with the War of Independence (1810) and the War of Reform (1855). Under this ideology, the Revolution provided a social and economic dimension to the fight, making it a transcendental reform for the nation (Florescano 2002: 380–84).
6. Since its foundation, Ex Teresa Arte Actual has been sponsored by the Instituto Nacional de Bellas Artes (INBA); the INBA is the government's cultural branch in charge of promoting artistic appreciation, education and production across the country (de la Garza 2008 online).

7. For Véjar Pérez-Rubio, environmental neocolonialism is the exportation of architecture from developed countries into developing ones. He argues that this often implies that the needs and particularities of developing countries are ignored in favour of paradigmatic architecture, which reflects the distinctive lifestyle, financial possibilities and technology advances of the centres of power (2007: 10–11).

Chapter 9 Post-conflict space

1. Commonly referred to as the phenomenon of time-space compression, post-modern temporal spatial relations and their impact are at the centre of much contemporary theory and analysis of space and place. Increased mobility through socio-economic migration occurs in parallel to rapid technological expansion that underpins a global networked system of people and capital. For in-depth reading through the concept of a global sense of place, refer to Massey, D. (1994) *Space, Place and Gender*, Cambridge: Polity Press and further in Harvey, D. (2007) *The Condition of Postmodernism: An Enquiry into the Origins of Cultural Change*, Oxford: Wiley-Blackwell.
2. The divergence of Protestant and Catholic historical narratives and their articulation through cultural memory are comprehensively analyzed in Dawson, G. (2010) *Making Peace with the Past*, Manchester University Press.
3. For a contrasting overview of the concepts of place and placelessness in humanist geography refer to Cresswell, T. (2006) *Place: A Short Introduction*, Oxford: Wiley-Blackwell, and in specific relation to Northern Ireland see Graham, B. (1997) *In Search of Ireland: A Cultural Geography*, London: Routledge.
4. These categories are not able to reflect the more complex origins of the respective community identities in Northern Ireland to which there originates a mono-cultural foundation. Historically, the Vikings, Normans, Scots and Anglo Saxons contribute to the notion of Irishness in parallel to the ancient Gaels. There has been politically involved Protestant support of Republicanism at various early historical intersects. The strength of 'Catholic-Gaelic' culture, in contrast to Protestant-Unionism, is founded in the connective harnessing of Gaelism, Catholicism and a cultural nationalism, extolled particularly through poetry, literature and music to achieve a consolidated narrative of identity and belonging. For a thorough analysis of Irish identity and concepts of nationhood refer to Kearney, R. (1997) *Postnationalist Ireland: Politics Culture Philosophy*, London: Routledge.
5. Since the 1998 Good Friday Agreement there has been an increase in the number of peace walls or peace lines, a total of 42 in Belfast in 2009 and a reported 99 barriers in 2012. The full report by the Belfast Interface Project is available at www.belfastinterfaceproject.org/. As noted in the text, there has been removal of some sectarian murals but they are equally a 'must see' landmark destination of city and 'Troubles' tours.
6. The challenge to humanist geographers' conception of place through phenomenology can be read in the work of American geographer Allan Pred (1984).

Chapter 10 Helen Chadwick

1. Chadwick's notebooks are held in the Helen Chadwick Collection, Henry Moore Institute, Leeds (hereafter HMI). They are cited here verbatim, and are referenced according to their archive accession number, for example, HMI 2003.19/E/5:99 – the last two digits give the sketchbook number (here, number 5) and page (here, page 99).

2. Chadwick asserts this many times. Here is just one example: 'A container can effect how we feel+think. If shape of instruments effects the musical note, so the shape of a space / thing must effect ourselves if part of us is exposed to / contained by it . . . Energy of body influenced by shape+containing structures, & vice versa . . . The geometrical space in which activity occurred modifies the person. result: physical→mental' (Chadwick, HMI 2003.19 / E / 5:99).
3. The text (re)produced as Chadwick 1992a is based on these lecture notes, and is also printed as Chadwick 1992b with different illustrations, and also clearly used as a base for other lectures given until 1995.
4. For this fuller discussion, see Walker 2012.
5. Chadwick refers to certain pages of this book in HMI 2003.19 / E / 6, and discusses the 'AH Reaction' in HMI 2003.19 / E / 7.
6. Nevertheless, she stressed her desire to ameliorate the problems she observed, rather than turn her back on them: she wanted to 'resolve reality not [take] flight from it [. . .] Change relationship with reality – affect it'. (HMI 2003.19 / E / 7:55) Her interest in this relationship between people and their environment remained every bit as strong, though this desire to 'change' and 'resolve' reality sought more oblique and nuanced approaches that she came to believe would provide more successful ways to counter the 'strength' of buildings. While this change is immediately manifest in her realized work (her next major project, *Ego Geometria Sum*, is frequently taken to mark the start of her 'mature' output) the range of issues just mentioned regarding *Model Institution* indicate several areas that she continued to consider.
7. This inscription of her own role as creative practitioner in a broader, institutionalised and legitimated area has strong parallels with the well-known work of Pierre Bourdieu. There is no evidence to suggest that Chadwick was aware of or had read Bourdieu, but his concept of *habitus*, fields of power and cultural capital, and their respective roles in maintaining the operation of culture and identity within society in ways that ensured the continuation of existing inequalities and power imbalances, would provide an interesting alternate framework within which to analyse *Model Institution* and so on. That is, unfortunately, beyond the scope of this essay.
8. In the *AMBIT* article, the actors are Silvia Ziranek and Brook Hoadley, in photographs of the performance taken at the ACME gallery, London, 1978: the piece was later performed by Chadwick and Philip Stanley, at the Spectro Gallery, Newcastle, 1979. In addition to the performance, there was a photo-romance style storyboard on panels behind the performance.
9. Her notebook from this period addresses passages from John Berger's *Ways of Seeing*, Edward T. Hall's *The Hidden Dimension: Man's Private and Public Use of Space*, Bernard Rudofsky's *The Unfashionable Human Body*, and Rudolf Broby-Johansen's *Body and Clothes: An Illustrated History of Costume.*
10. Koestler insisted on using the term 'Philosophy of Nature' rather than 'science' because the former carried so many more overtones: see Koestler 1982 [1959]: 9.

Chapter 11 Identity, culture, and urbanism

1. The argument and material included in this paper are an outgrowth of earlier arguments in AlSayyad (ed.) *Forms of Dominance* (1993) and more recently in Avermaete, T., Karakayali, S. and Von Osten, M. (eds) (2010) *Colonial Modern: Aesthetics of the Past-Rebellions of the Future*, London: Black Dog Publishing.

2. Many conceptions of globalization have been developed in the social sciences or are rooted in economic theories. This paper mainly builds on the discourse in the field of cultural studies.
3. Many of the lessons below are based on the work of Anthony King, Stuart Hall, Roland Robertson, Immanuel Wallenstein, Ulf Hannerz, and Janet Wolff, presented at a symposium held at the State University of New York, Binghamton, in 1989, and appearing in the edited volume, *Culture, Globalization, and the World System* (King 1991).
4. The term 'virtual reality' was introduced in computer science in the 1980s to designate nearly exact simulations of audiovisual experience. The field is now growing to include simulations of holistic environments.

Chapter 12 Space of representational negotiations

1. Portions of this chapter have appeared previously in *Architecture of Thought* (Piotrowski 2011).
2. The International Research and Exchanges Board supported this part of my research. Mgr. Zygmunt Nasalski, director of the Muzeum Lubelskie, generously granted the permission to study and photograph the chapel.
3. As such, it was the object of intense missionary work undertaken by the Catholics from the West and the Orthodox from the south. For discussion of similar issues, see also chapter 3 in *The Reformation in Lithuania: Religious Fluctuations in the Sixteenth Century* (Musteikis 1988).
4. Lublin was the town where Jagiełło finalized his political negotiations in 1386 and where, in the same year, he stopped on his way to Kraków to be crowned. See *Polska Jana Długosza* (Samsonowicz 1984: 180). When the alliance matured almost two centuries later, in 1569, the final document legalizing the Commonwealth of the Kingdom of Poland and the Grand Duchy of Lithuania was also signed in Lublin and religious celebrations took place in the Holy Trinity Chapel (Durakiewicz 1995).
5. Although the history of the building remains insufficiently explored recent restoration works have uncovered that the building already had shown traces of use when walls had been covered with paintings (Durakiewicz 1995).
6. Anna Różycka-Bryzek (1992: 338) acknowledges this general shift in this and other Polish churches where Gothic architecture coexisted with elements of Russian Byzantine paintings.
7. Tadeusz Trajdos (1979: 27) asserts that such political efforts culminated in 1415, according to him the date when the chapel paintings were completed, when Jagiełło convinced Grzegorz Camblak (Григорий Цамблак), the head of Eastern Orthodoxy in the Duchy of Halicz-Vladimir, to accept the administration of the Roman pope.
8. According to Tadeusz Bernatowicz, who references Jan Ambroży Wadowski (Bernatowicz, 2000: 17), Jan Cangier constructed the gable in 1627–29 CE (or possibly in 1635–42 CE). The initial shape of the Gothic gable still appears in an engraving of 1618 CE.
9. It is noteworthy that between 1596 and 1603, three new altars were made for the interior of the chapel. Later moved to the Church of St. Adalbert (Kościół Św. Wojciecha) located nearby, they are among the oldest, most exquisitely crafted, and best preserved examples of northern European Mannerism in that part of Poland.
10. According to Aleksander Kossowski (1933: 17), this multicultural mix was acknowledged by Venetian nuncio Hieronim Lippomano and Alexander Guagnini.

11. Faustus Socinus, the leader of the group, was born and educated in Italy and arrived in Poland in 1579 CE.
12. According to Stanisław Załęski (1908: 3), this term was used, for example, by nuncio Ruggieri in 1568 CE.
13. One of the brightest chapters in the history of the Commonwealth is the Compact of Warsaw. While in other countries of Europe, Roman Catholicism and the Protestant religions struggled for domination in bloody battles and persecutions, in 1573 CE – five months after and partly in response to the Massacre of St. Bartholomew's Day in France – religious tolerance was legally asserted in Poland and Lithuania by the Compact of Warsaw, in Latin called *Pax Dissidentium* (The Peace of Those Who Differ).

Chapter 13 Colour and tradition

1. The Mexican Revolution (1910–1920) ended the thirty-year dictatorship of Porfirio Díaz and ushered in a period of agrarian reform and social renewal.
2. The Spanish word *mestizo* is used to refer to people of mixed blood.
3. Both of these trends were present in all of Latin America and the bibliography on the subject is extensive. See for example: Aracy Amaral (1994).
4. For example, Manuel Amabilis argued '. . . we must recognize that the arts of our grandfathers have a vitality, characteristics so genuine that, in spite of the abandonment of centuries, even now they awaken interest in the foreigner so that our contemporary art and artists are relegated, almost eliminated in the face of the admiration arising from Mexican antiquities.' (Ríos Garza *et al.* 2001: 41)
5. The introductory texts appear in both English and Spanish; the descriptions of selected works are in English only.
6. For example, in reference to a house designed by Guillermo Rossel and Lorenzo Carrasco he stated 'The terra-cotta exterior, the natural tones of wood and stone, and the large window expanses are in refreshing contrast to adjacent Spanish Colonial residences [. . .]' (Myers 1952: 116).
7. Verna Cook Shipway, the main author, was an architect who had studied at Columbia University in 1919 and led a very successful career in New York until 1940 when she retired.
8. All of the books were successful with ample distribution on both sides of the border (Dudley 1968).
9. The book was published by the same company as the Shipway's books.
10. The name was changed to AM Editores in 2003.

Chapter 14 To be for (an)*other*

1. The original text is written in Spanish. This translation is by the author.
2. For Walter Mignolo 'coloniality' is the rational structuring of control or the logistics of domination. Colonialism, on the other hand, is defined as imperial domination that has taken place over a specific time-space. Mignolo uses 'coloniality' to reference power structures still at work even after independence or decolonization.
3. The original text is written in Spanish. This translation is by the author.
4. According to the proposal, the Government of Puerto Rico would invest $7.5 million while Hilton would put up $500,000. The Government of Puerto Rico would own the hotel while the Hilton Company would lease and administer the property. As revenue agreement, the Hilton Company would pay to PRIDCO two-thirds of the brute earnings, plus five per cent of the cost of machinery and equipment (Maldonado 1997: 123).

5. The Schimmelpfenning, Ruiz and González entry has apparently been lost.
6. Henry Klumb graduated from the *Sataalische Bauschule* of Cologne in 1926. In 1929 he joined Frank Lloyd Wright's studio, eventually becoming chief draftsman and one of the original Taliesin Fellows. After leaving Wright in 1933, Klumb worked in several states as architect and planner until his 1944 arrival in Puerto Rico to join the Committee on Design of Public Work. Refer to Henry Klumb Collection at the Architecture and Construction Archives at the University of Puerto Rico.
7. That view was also present in the house in Río Piedras that became Klumb's tropical laboratory.
8. This stems directly from the design philosophy of Frank Lloyd Wright, Klumb's mentor. Perfectly linking with Wright's attitude in the projects for Arizona, especially San Marcos in the Desert, it seems Klumb intended for his hotel to integrate with the context in a less literal way.
9. With the added elevation of the guest room in a multi-storey building, the architect allowed for a purely modern visual experience – such as that afforded through the birds-eye perspective or from the airplane. Although conversely, the viewing from above has been historically linked to a discourse of control.
10. Although Ferrer claimed that they did not know of Le Corbusier at the time, conversations with architect Juan Marqués Mera, who has researched the architectural firm extensively, confirmed that, for inspiration, Toro and Ferrer looked at Alberto Sartori's *Gli Elementi dell'Architettura Funcianale* (1932) – a book in Alberto Toro's library. A review of that publication verifies not only that Le Corbusier wrote the preface to the edition, but also that his work is shown in 26 pages as the French representative of modern architecture in countries of Latin extraction.
11. Bruno Stagno (2001: 75) suggests that while rational Cartesian thought is linear, tropical thought, translated as 'I feel therefore I am', is spatial.
12. A clear example is found in Muñoz Marín's *Gotkin Lectures* of 1959.
13. The argument is amplified by the research of James Dietz (1986: 133) who points out that due to the capitalist production system, Puerto Rico distanced itself from Latin American and Caribbean societies. The economist also argues that eventually, the Hispanic past was made invisible due to the strong effects that the capitalist current had in Puerto Rican politics, economy, and culture. In that same vein, Ricardo Alegría, first director of the Institute of Puerto Rican Culture, explained that when the proposal for the *San Juan Historic Zone* project started in 1955, some business sectors strongly opposed it as they believed Spanish colonial San Juan should become a mini New York (Reina Pérez 2003: 112). That way of thinking, Alegría stressed, certainly supported the negation of a Hispanic heritage.

Chapter 15 The mirror of territorial identity in Singapore professional architecture 1923–1969

1. This cultural duality is part of a broader discourse that deserves fuller explication that the author will give elsewhere; it is sufficient for this chapter to recognize the existence of this whilst a particular reciprocal duality of the professional architectural institute, its journal and status of territorial identity are herewith examined.
2. In the colonial era, Singapore, Penang and Malacca were part of the Straits Settlements until 1948 when the Federation of Malaya States incorporated Penang and Malacca, leaving Singapore a Crown colony. From 1955 Singapore worked with the northern hinterland's Federation to seek independence from the British, at a time when Singapore's status in relation to the Federation was an open negotiation without resolution.

The Federation of Malaya gained independence in 1957. Singapore received its status of self-governance from colonial Britain in 1958 but formal announcement was delayed until 1959. In August 1963, Singapore gained independence from the British and became a part of the Malayan Federation for a short period, before Singapore's ejection and eventual republican independence in 1965. The territory of identity claimed within the political realm reflects a peculiar competitive duality, extending even to language. The word '*merdeka*' meaning 'independence' was the rallying cry for Malayan independence from the British; it was a word that was used in the independence movement in Singapore in the 1950s (Drysdale 1984). This term is now totally associated with Malaysian independence and is no longer used in Singapore to mark national independence, when it once appeared in the Independence movement documents. *Petir*, the People's Action Party Journal, published an article in 1958 titled 'The New Phase after *Merdeka* – Our Task and Policy', a strategy for a multicultural view in a merger with Malaya, that was required reading for all PAP candidates in the 1959 elections (Bloodworth 2005). Yet when Singapore became independent, its national language was officially 'Malay'. When Singapore sought to join the Malayan Federation, the term used was 'merger', belying its actual proportionate size. There are many other aspects of life, entities, and assets that exhibit this duality in the claim of territorial identity, such as kinship across territories, land and resource ownership, and even claims to historical legacies. The Malaysia-Singapore Airlines oddly came into being in 1966, a year after separation, as a jointly owned asset of the governments of Malaysia and Singapore, separating into two independent national carriers only in 1972. A currency union operated from 1938 until 1967 when Malaysia, Singapore and Brunei began independent currencies. These were interchangeable until 1973 when Malaysia opted out of the 'Interchangeability Agreement' it held with Brunei and Singapore (MAS 2012). The symbols of the two territories are feline: the tiger is a genuine Malayan species, and in contra-distinction to this, Singapore celebrates a competing feline, the lion.

3. Campbell Oman was a partner at Westerhout and Oman, and a president of SSA from 1924–25.
4. The society's committee comprised primarily expatriates, but featured a few locals such as the partners at Chung and Wong.
5. In what appears to be a verbatim transcription of the meeting, Meadows, the immediate past president, opposes Wilson, Santry and others for running a select club. Santry and Wilson blame apathy. It is clear from the minutes that the SSA was not representative of all (JSSA Aug 1925).
6. It may be striking to observe that Britain had not yet added registration of architects to law at the time, as the debates were on-going, and it was a UK Parliament Act in 1931 that saw the establishment of the Architects' Registration Council of the United Kingdom (ARCUK, precursor the to the present day ARB).
7. Ho Kwong Yew, a native bred architect, would later become a celebrated professional talent (see note 15).
8. *Singha* is Sanskrit for lion.
9. A record of the original formation of the KL branch, pre-cursor to FSMA, is noted in this issue. The differences in ordinances gave the SSA and IAM some concern (JSSAI Feb 1930: 2–3).
10. Via PETA, the FSMA is incorrect to refer to the formation of the IAM as the point of their origin, the SSA/SSAI preceded this and the IAM is its continuation, but of course as competing entity it would not concede this. PAM incorrectly carries the same claims of the FSMA. PAM claims the IAM originated in 1920, which is not correct.

11. The JSSAI editorial 'comment' page is always signed 'H.S.', viz. Honorary Secretary. When comments are made in the JSSA, Wilson signed as 'ed.'
12. The SSA started on an unsure footing, and Meadows who was president did not seem to envisage a stake and left the colony in 1925. Wilson and others openly attacked Meadows for his lack of commitment in a fully transcribed SSA meeting (JSSA Aug 1925: 3).
13. *Townplanner* was in all probability James Milner Fraser, a Scot who arrived via London in 1928 to work for the Singapore Improvement Trust, and who was once a student of Alexander Gordon at Aberdeen. Fraser wrote an historical review of the SIT's work in the region in 1953, and during the years of 1929–33, he was the municipal officer for planning. When featured in TMA later, he talked about town planning in the same critical tone (TMA Oct 1935: 228–29).
14. Something peculiar happened here which requires deeper investigation, and is beyond the remit of this chapter.
15. Ho was tragically killed by the Japanese during WWII.
16. The Federation of Malaya Society of Architects published a journal commencing 1955, some 4 years after the QJIAM, and some 6 years after its inception.
17. Hancock was editor but not secretary of the institute, a structural practice differing from the JSSA, JSSAI, and JIAM. He was to remain in this role under successive administrations. Hancock later published a book titled *Coleman's Singapore*. Hancock served on the IAM and FSMA councils simultaneously.
18. George Anthony Atkinson's paper 'Building in the tropics' first published in the RIBA Journal in 1948 was re-published in full. A conference on the tropics took place in London in 1953, and climatic response was a topical subject (Conference 1953). The clearest example of this global phenomenon can be seen in the Catholic Church's adaptations to climates (*Liturgical Arts* 1953).
19. A number of historical resources will describe the uncertainty and bloodshed. See for example Bloodworth, 2005 and Drysdale, 1984.
20. Victor Chew was one of the founding members of the SMA, and later principal of Kumpulan Akitek, a practice with interests in Singapore and Malaysia.
21. Lim Chong Keat, originally from Penang, operated the Malayan Architects Co-Partnership across Singapore and KL. Tay Kheng Soon, who worked at the MAC, reported that 'his face was black' the day Singapore was ejected in August 1965 (Tay 1987). Lim continues to live in Penang and operated for many years on both sides of the peninsular after the separation. Tay has stated in conversation that his own 'Malayanism' was inspired by Lim.
22. SIAJ was published monthly until 1996, when its name changed to *Singapore Architect*, commencing with issue 189, which continued the numbering of volumes.
23. A book was published in 1981 titled *Rumah*, but this is clearly not volume 10 and is something else entirely.
24. I have written about the issues of identity construction and sedimentation elsewhere (Quek 2012a: editorial).

Bibliography

Abbinnett, R. (2004) *Culture and Identity: Critical Theories*, London: Sage.

Abu-Lughod, J. (1991) 'Going Beyond Global Babble', in A. King (ed.) *Culture, Globalization, and the World System*, London: Macmillan, pp. 131–138.

Adriá, M. (1996) *México 90's: Una arquitectura contemporánea*, México: Editorial Gustavo Gili.

Akbar, J. (1988) *Crisis in the Built Environment*, Singapore: Concept Media.

Akcan, E. (2008, in Turkish) *Çeviride Modern Olan: Şehir ve Konutta Türk-Alman İli şkileri*, Istanbul: YKY – YapıKredi Yayınları.

Al-Khalil, S. (1991) *The Monument: Art, Vulgarity and Responsibility in Iraq*, Berkeley: University of California Press.

Albu, E. (2005) 'Imperial Geography and the Medieval Peutinger Map', *Imago Mundi*, vol. 57, Part 2, pp. 136–48.

AlDallal, E., AlWaer, H. and Bandyopadhyay, S. (in press) *Site and Composition: Design Strategies in Architecture and Urbanism*. London: Routledge.

Alföldi, A. (1949) 'The Moral Barrier on Rhine and Danube', in E. Birley (ed.) (1952) *The Congress of Rome Frontier Studies*, Durham: University of Durham, pp. 1–16.

AlSayyad, N. (1992) 'Urbanism and the Dominance Equation', in N. AlSayyad (ed.) *Forms of Dominance: On the Architecture and Urbanism of the Colonial Enterprise*, London: Avebury, pp. 1–26.

AlSayyad, N. (ed.) (2003) *The End of Tradition?* London: Routledge.

AlSayyad, N. and Bourdier, J.P. (1989) *Dwellings, Settlements, and Tradition*, New York: University Press of America.

Amábilis, M. (2003) *Arquitectura Nacional*, Mexico City: INBA.

Amaral, A. (coord) (1994) *Arquitectura Neo-colonial, América Latina, Caribe, Estados Unidos*, Sao Paolo: Memoria and Fondo de Cultura Económica.

Ambasz, E. (1976) *The Architecture of Luis Barragan*, New York: The Museum of Modern Art.

Andermann, J. and Arnold-de Simine, S. (2012) 'Introduction: Memory, Community and the New Museum', *Theory Culture Society*, 29, 1: 3–13.

Arai, A. T. (2001 [1952]) *Caminos para una arquitectura mexicana*, Mexico City: INBA.

Architects' Ordinance no. 48 of 1951, (18 October 1951), Malaya

Architectural Forum (1950) 'Spectacular Luxury in the Caribbean: The Caribe Hilton Hotel at San Juan, Puerto Rico', 92/3 (March 1950): 97–105.

Architectural Review (June 1960), London.

Armitage, J. (ed.) (2000) *Paul Virilio: From Modernism to Hypermodernism and Beyond*, London: Sage Publications.

Arnold-de Simine, S. (2005) *Memory Traces: 1989 and the Question of German Cultural Identity*, Bern: Peter Lang.

Attenbrow, V (2005) 'Aboriginal Placenames In Sydney Harbour, New South Wales, Australia: A Dual Naming Project', paper presented at the forum UNESCO University and

Heritage 10th International Seminar, *Cultural Landscapes in the 21st Century*, Newcastle-uponTyne, 11–16 April 2005. Revised: July 2006 Online. Available at www.ncl.ac.uk/unescolandscapes/files/ATTENBROWVal.pdf> (accessed 13 January 2012).

Auge, M. (1995) *Non-Places: Introduction to an Anthropology of Supermodernity*, trans. J. Howe, London: Verso.

Australian Human Rights Commission (1997) *Bringing Them Home: the 'Stolen Children' Report.* Available at www.hreoc.gov.au/social_justice/bth_report/index.html (accessed 3 June, 2012).

Ayer, E. H. (1994) *The United States Holocaust Memorial Museum: America Keeps the Memory Alive*, New York and Ontario: Dillon Press.

Baker, L. (2007) 'On Being an Architect'. Online. Available at lauriebaker.net/work/work/on-being-an-architect.html>: unpaginated (accessed 3 September 2009).

Bandyopadhyay, S. and Jackson, I. (2007) *The Collection, the Ruin and the Theatre: Architecture, Sculpture and Landscape in Nek Chand's Rock Garden, Chandigarh*, Liverpool: Liverpool University Press.

Barber, B. (1995) *Jihad vs. McWorld*, New York: Ballantine Books.

Barnard, M. (1962) *A History of Australia*, Sydney: Angus and Robertson.

Beacham, H. (1969) *The Architecture of Mexico: Yesterday and Today*, New York: Architectural Book Publishing Co.

Beard, M. (2007) *The Roman Triumph*, Cambridge, Mass: Belknap Press, Harvard University Press.

Bell, C. (1992) *Ritual Theory, Ritual Practice*, New York: Oxford University Press.

Bell, V. (2011) *Contemporary Art and Transitional Justice in Northern Ireland: The Consolation of Form* [Article]: Goldsmiths Research Online. Available at eprints.gold.ac.uk/3740/ (accessed December 2011).

Benjamin, W. (1968 [1935]) 'The Work of Art in the Age of Mechanical Reproduction', in *Illuminations*, Introduction by H. Arendt, New York: Schocken Books.

Bennett, T. (1995) *The Birth of the Museum: History, Theory, Politics*. London and New York: Routledge.

Bernatowicz, T. (2000) 'Rola Lublina w architekturze sakralnej Wielkiego Księstwa Litewskiego', in J. Lilejko (ed.) *Sztuka ziem wschodnich Rzeczypospolitej XVI–XVIII w*, Lublin: Towarzystwo Naukowe KUL.

Berque, A. (1997) *Japan: Cities and Social Bonds*, trans. C. Turner, Yelverton Manor: Pilkington Press.

Bevan, R. (2006) *The Destruction of Memory. Architecture at War*, London: Reaktion Books.

Bhabha, H. K. (1990) 'DissemiNation: Time, Narrative, and the Margins of the Modern Nation', in H. Bhabha (ed.) *Nation and Narration*, London: Routledge.

Bhabha, H. K. (1994) *The Location of Culture*, London: Routledge.

Bhabha, H. K. (2010 [1994]) *The Location of Culture*, London & New York: Routledge.

Bhatia, G. (1994) *Laurie Baker: Life, Works and Writings*, New Delhi: Penguin.

Bloodworth, D. (2005) *The Tiger and the Trojan Horse*, Singapore: Times Editions/Marshall Cavendish.

Bloomer, K. and Moore, C. (1977) *Body, Memory and Architecture*, New Haven: Yale University Press.

Boeck, F. (1998) 'The rootedness of Trees: Place as Cultural and Natural Texture in Rural Southwest Congo', in N. Lovell (ed.) *Locality and Belonging*, London: Routledge.

Bognar, B. (1988) 'Archaeology of a Fragmented Landscape', in B. Bognar (ed.) *Japanese Architecture*, London: Academy Group, pp. 15–25.

Bognar, B. (1990) 'Archaeology of a Fragmented Landscape: the New Avant-garde of Urban Architecture in Japan', in B. Bognar (ed.) *The New Japanese Architecture*, New York: Rizzoli.

Born, E. (1937) *The New Architecture in Mexico*, New York: The Architectural Record.

Bourdieu, P. (1977) *Outline of a Theory of Practice*, trans. R. Nice, Cambridge: Cambridge University Press.

Bourdieu, P. (2005) 'Habitus', in J. Hillier and E. Rooksby (eds) *Habitus: A Sense of Place*, Aldershot: Ashgate.

Boyer, M.C. (1996) *The City of Collective Memory: Its Historical Imagery and Architectural Entertainments*, Cambridge, Mass: MIT Press.

Boyer, M.C. (2012) 'Collective Memory Under Siege: The Case of "Heritage Terrorism"', in C. G. Crysler, S. Cairns and H. Heynen (eds) *The Sage Handbook of Architectural Theory*, Los Angeles, London, New Delhi, Singapore, Washington DC: SAGE Publications, pp. 325–339.

Bozdoğan, S., Özkan, S., and Yenal, E. (1987) *Sedad Eldem, Architect in Turkey*, New York: Aperture.

Breslin, L. (1988) 'From the Savage to the Nomad: Critical Interventions in Contemporary Japanese Architecture', in B. Bognar (ed.) *Japanese Architecture*, London: Academy Group.

Bridge, G. (2005) *Reason in the City of Difference: Pragmatism, Communicative Action and Contemporary Urbanism*, London: Routledge.

Bristol, K. (1991) 'Pruitt-Igoe Myth', *Journal of Architectural Education*, pp. 163–171.

Brown, R. (2008) 'Redefining the Edge', in N. AlSayyad (ed.) *Reinventing Traditions for the Modern World: Traditional Dwellings and Settlements Working Paper Series*, 211: 1–24; Berkeley: IASTE.

Brown, R. (2011) 'The Elusiveness of Culture', in P. Beacock et al. (eds) *Engaging in Architectural Education*, London: London Metropolitan University, pp. 49–53.

Brown, R. and Maudlin, D. (2012) 'Concepts of the Vernacular', in S. Cairns, G. Crysler and H. Heynen (eds) *Handbook of Architectural Theory*, London: Sage Publications.

Bruno, G. (2007) *Public Intimacy: Architecture and the Visual Arts*, London and Cambridge, Mass: MIT Press.

Buczkowa, I. (1983) *Kościół Świętej Trójcy na Zamku Lubelskim*, Lublin: Muzeum Okręgowe.

Burdett, R. and Sudjic, D. (eds) (2007) *The Endless City: The Urban Age Project by the London School of Economics and Deutsche Bank's Alfred Herrhausen Society*, London: Phaidon.

Burian, E, (ed.) (1997) *Modernity and the Architecture of Mexico*, Austin, Texas: University of Texas Press.

Butler, J. (1999) 'Performativity's Social Magic', in R. Shusterman (ed.) *Bourdieu: A Critical Reader*, Oxford: Blackwell.

Calderón, V. (2011) 'Puerto Rico es EEUU, pero no todos lo tienen claro', *El País*. Online. Available at http://el pais.com/diario/2011/01/21/internacional/129554413_850215.html (accessed 21 May 2012)

Caponetto, S. (1999) *The Protestant Reformation in Sixteenth-Century Italy*, trans. A.C. Tedeschi and J. Tedeschi, Kirksville, MO: Thomas Jefferson University Press.

Carl, P. (2003) 'The Tower of Shadows', in *Le Corbusier and the Architecture of Reinvention*, London: Architectural Association.

Carrier, P. (2005) *Holocaust Monuments and National Memory: France and Germany since 1989*, New York: Berghahn Books.

Casey, E. S. (1993) *Getting Back into Place: Toward a Renewed Understanding of the Place-World*, Bloomington: Indiana University.

Castells, M. (1992) 'The World Has Changed: Can Planning Change?', *Landscape and Urban Planning*, 22: 29–40.
Chadwick, H. (1979) 'Objects' in *AMBIT*, 81: 17–22.
Chadwick, H. (1989) 'Soliloquy to Flesh,' in *Enfleshings*, London and New York: Secker and Warberg.
Chadwick, H. (1990) Preparatory notes for a lecture given at the Symposium for Photography in Gratz in October of that year. (HMI Box 19: lecture notes).
Chadwick, H. (1992a) 'Withdrawal: Object, Sign, Commodity', *Architecture.Space.Painting, Journal of Philosophy and the Visual Arts*, London: Academy Group, pp. 68–73.
Chadwick, H. (1992b) 'Withdrawal: Object, Sign, Commodity' in *Camera Austria*, 37: 4–10.
Chew, V. C. A. (undated) *Urban Planning in Singapore*, National Archives, Singapore, Oral History Centre, ascension number 002786, Reel / Disc No.13 of 30.
Claridge, A. (with contributions by J. Toms and T. Cubberley) (1998) *Rome: An Oxford Archaeological Guide*, Oxford: Oxford University Press.
Clendinnen, I. (2003) *Dancing with Strangers*, Melbourne: Text Publishing.
Coarelli, F. (2000) *The Column of Trajan*, Rome: Editore Colombo (in collaboration with the German Archaeological Institute).
Collins, D. (1798–1802; B. H. Fletcher ed., 1975) *An Account of the English Colony in New South Wales, with Remarks on the Dispositions, Customs, Manners, etc. of the Native Inhabitants of that Country*, Sydney: Reed in association with the Royal Historical Society. First published, London: T. Cadell and W. Davis, 1798–1802.
Compañía de Fomento Industrial de Puerto Rico (1949) *Informe anual, año fiscal 1948–1949*, San Juan. PRIDCO.
Conference on Tropical Architecture: A Report on the Proceedings of the Conference, held at University College, London, March 1953.
Corcoran, F. (2002) 'The Political Instrumentality of Cultural Memory: A Case Study of Ireland', *The Public* vol. 9, 3, 49–64.
Cornell, L. (1997) 'House Architecture and Family Form: On the Origin of Vernacular Traditions in Early Modern Japan', *Traditional Settlements and Dwellings Review*, VIII (II): 21–31, Berkeley: IASTE.
Cosgrove, D. (1992) 'Mapping new Worlds: Culture and Cartography in Sixteenth Century Venice', *Imago Mundi: The International Journal for the History of Cartography*, vol. 44, Issue 1: 65–89.
Cosio Villegas, D. *et al.* (eds) (2003) *Historia mínima de México*, México: El Colegio de México.
Cresswell, T. (2006 [2004]) *Place: A Short Introduction*, Oxford: Wiley-Blackwell.
Cubitt, G. (2007) *History and Memory*, Manchester: Manchester University Press.
Curtis, W.J.R. (1996) *Modern Architecture since 1900*, London and New York: Phaidon.
d'Harnoncourt, R. (1931) *Mexicana: A Book of Pictures*, New York: Alfred A. Knopf.
Davies, G.A.T. (1920) 'Topography and the Trajan Column', *Journal of Roman Studies*, vol. 10: 1–28.
Davies, P. J. E. (1997) 'The Politics of Perpetuation: Trajan's Column and the Art of Commemoration', *American Journal of Archaeology* 101: 41–65.
Davies, P. J. E. (2000) *Death and the Emperor: Roman Imperial Funerary Monuments from Augustus to Marcus Aurelius*, Cambridge: Cambridge University Press.
Dawson, G. (2010) *Making Peace with the Past*, Manchester: Manchester University Press.
de Anda Alanís, E. (1998) '1920–1970: cincuenta años de arquitectura en la Ciudad de México', in E. de Anda Alanís (ed.) *Ciudad de México: Arquitectura 1921–1970*, Mexico: Gobierno de la Ciudad de México and Junta de Andalucía, pp. 25–54.

de Anda Alanís, E. (2005) *Una mirada a la arquitectura mexicana del siglo XX (Diez ensayos)*, México: CONACULTA.

de Anda Alanís, E. (2006) *Historia de la arquitectura mexicana*, Barcelona: Gustavo Gili.

de la Garza, C. (2008) 'Ex Teresa: Hacia los 15 años de performance en la ciudad de México', *Arteamerica*. Online. Available at http://www.arteamerica.cu/14/dossier/delagarza.htm (accessed 18 July 2009).

Deane, D. (2009) 'The Recessive Artefact', in D. Deane (ed.), *Lateralisms*, Nottingham: Nottingham University.

Delpar, H. (1992) *The Enormous Vogue of Things Mexican: Cultural Relations between the United States and Mexico, 1920–1935*, Tuscaloosa: University of Alabama Press.

Dietz, J. (1986) *Economic History of Puerto Rico: Institutional Changes and Capitalist Development*, New Jersey: Princeton University Press.

Digital Daijisen (Dictionary), Tokyo: Shogakukan (2011a) 'Shikii wo matagu'. Available at http://kotobank.jp/word/shikii wo matagu (accessed 12 May 2012).

Digital Daijisen (Dictionary), Tokyo: Shogakukan (2011b) 'Genkan'. Available at http://kotobank.jp/word/genkan (accessed 12 May 2012).

Digital Daijisen (Dictionary), Tokyo: Shogakukan (2011c) 'Agarigamachi'. Available at http://kotobank.jp/word/genkan (accessed 12 May 2012).

Digital Daijisen (Dictionary), Tokyo: Shogakukan (2011d) 'I(tsu)tekimasu'. Available at http://kotobank.jp/word/i(tsu)tekimasu (accessed 12 May 2012).

Digital Daijisen (Dictionary), Tokyo: Shogakukan (2011e) 'I(xtu)tera(xtu)shai'. Available at http://kotobank.jp/word/i(xtu)tera(xtu)shai (accessed 12 May 2012).

Digital Daijisen (Dictionary), Tokyo: Shogakukan (2011f) 'Tadaima'. Available at kotobank.jp/word/tadaima (accessed 12 May 2012).

Digital Daijisen (Dictionary), Tokyo: Shogakukan (2011g) 'Okaerinasai'. Available at http://kotobank.jp/word/okaerinasai (accessed 12 May 2012).

Dio Cassius (1924) *Roman History, Volume VII, Books 56–60*, trans. E. Cary and H. B. Foster, Cambridge Mass: Harvard University Press, Loeb Classical Library.

Dovey, K. (2008 [1999]) *Framing Places: Mediating Power in Built Form*, London: Routledge.

Dresler-Hawke, J. H. and Liu, E. (2006) 'Collective Shame and the Positioning of German National Identity', *Psicología Política*, 32: 131–153. Available at http://www.uv.es/garzon/psicologia%20politica/N32-7.pdf (accessed 6 February 2013).

Drew, P. (2002 [1995]) *Sydney Opera House: Jørn Utzon/Philip DrewI*, London: Phaidon Press.

Drysdale, J. (1984) *Singapore: Struggle for Success*, Singapore: Times Editions/Marshall Cavendish.

Dudley, L. (1968) 'Husband and Wife Write on Mexico', *Evening Tribune*, San Diego, 30 September.

Duffy, P.J. (1997) 'Writing Ireland: Literature and Art in the Representation of Irish Place', in B. Graham, (ed.) *In Search of Ireland: A Cultural Geography*, London: Routledge.

Durakiewicz, K. (1995) 'Kaplica', *Na Przykład*, no. 27.

Durham, J. (2004) 'A Thousand Beautiful Things: Nikos Papastergiadis, Jimmie Durham and Maria-Theresa Alves', in B. French, A. Geczy and N. Tsoutas (eds) *Criticism+Engagement+Thought: Biennale of Sydney 2004*, Sydney: Artspace.

Eagleton, T. (1981) *Walter Benjamin or Towards a Revolutionary Criticism*, London: Verso.

Edkins, J. (2003) *Trauma and the Memory of Politics*, Cambridge: Cambridge University Press.

Eisenman, P. (1971) 'From Object to Relationship II: Giuseppe Terragni, Casa Giuliani Frigerio', *Perspecta* 13, The Yale Architectural Journal.

El-Ojeili, C. and Hayden, P (2006) *Critical Theories of Globalization*, Basingstoke: Palgrave Macmillan.

Ellem, L. (1980) 'Utzon's Sydney Opera House', in A. Bradley, T. Smith and B. Smith (eds.) *Australian Art and Architecture: Essays Presented to Bernard Smith*, Melbourne: Oxford University Press.

Espinosa López, E. (2007) *Ciudad de México. Compendio cronológico de su desarrollo urbano 1521–2000*, México: Instituto Politécnico Nacional.

Farnell, B. (1995) *Do You See What I Mean?*, Austin: University of Texas.

Florescano, E. (2002) *Historia de las historias de la nación Mexicana*, México: Taurus.

Fortuño, L. (2011) 'Puerto Rico punto de encuentro', *Diálogo digital*. Available at http://www.dialogodigital.com/index.php/Puerto-Rico-punto-de-encuentro . . . -discurso-de-Luis-Fortuno-en-U.-de-Vallalodid.html (accessed 21 May 2012).

Foster, H. (1985) *Postmodern Culture*, London: Pluto Press.

Foucault, M. (1986, Miskowiec, J. tr.) 'Of Other Spaces', *Diacritics* 16(1): 22–27.

Frampton, K. (1985) 'Towards a Critical Regionalism: Six Points for an Architecture of Resistence', in H. Foster (ed.) *Postmodern Culture*, London: Pluto Press.

Frampton, K. (1998) 'Introduction', in K. Frampton, A. Spector and L. R. Rosman (eds.) (1998) *Technology, Place and Architecture: The Jerusalem Seminar in Architecture*, New York: Rizzoli, 12–15.

Frascari, M. (1996) 'The Tell-The-Tale Detail', in K. Nesbitt (ed.) *Theorizing a New Agenda for Architecture: An Anthology of Architectural Theory 1965–1995*, New York: Princeton, 500–514.

Freud, S. (1965 [1900]) *The Interpretation of Dreams*, trans. and ed. J. Strachey, New York: Avon Books.

Freud, S. (1994 [1930]) *Civilization and Its Discontents*, trans. J. Riviere, New York: Dover Publications.

Fullbrook, M. (1999) *German National Identity after the Holocaust*, Oxford: Blackwell.

Gadamer, H-G. (1996) *The Enigma of Health*, Stanford, California: Stanford University Press.

Gamas Torruco, J. (2012) *México y la Constitución de Cádiz*, Mexico: Secretaría de Governación, UNAM and Museo de las Constituciones.

Garma-Montiel, G., Bandyopadhyay, S. and Purssord, L. (2009) *The Multiple Faces of Identity*, Nottingham: Nottingham Trent University.

Geographical Names Board (2012) *Bennelong Point*. Available at http://www.gnb.nsw.gov.au/name_search/extract?id=ujKqFxWA (accessed 13 January 2012).

Gibson, J. (1977) 'The Theory of Affordances', in R. Shaw and J. Bransford (eds) *Perceiving, Acting, Knowing: Towards and Ecological Psychology*, Hillsdale: Lawrence Erlbaum.

Gibson, J. (1979) *The Ecological Approach to Visual Perception*, Boston: Houghton Mifflin.

Giddens, A. (1990) *The Consequences of Modernity*, Cambridge: Polity; also published in Stanford: Stanford University Press.

Giunta, A. (2005) 'Misión imposible: Nelson Rockefeller y la cruzada del internacionalismo artístico', in R. Salvatore (ed.) *Culturas imperiales: Experiencia y representación en América, Asia y África*, Argentina: Viterbo Editora, pp. 186–211.

Google Inc (2012) *Google Maps*. Available at http://maps.google.com.au/ (accessed 3 January 2012).

Grafton, A. (1995) *New Worlds, Ancient Texts*, Cambridge, Mass: The Belknap Press of Harvard University Press.

Graham, B. (1994) 'No place of the Mind: Contested Protestant Representations of Ulster', *Ecumene* 1:3. 257–281.

Graham, B. (1997) *In Search of Ireland: A Cultural Geography*, London: Routledge.

Halbwachs, M. (1992 [1941]) *On Collective Memory*, ed., trans. and Introduction by L. A. Coser, Chicago and London: University of Chicago Press.

Hall, S. (1991) 'Old and New Identities, Old and New Ethnicities', in A. King (ed.) *Culture, Globalization, and the World System*, London: Macmillan, pp. 41–68.

Hardt, M. and Negri, A. (2000) *Empire*, Cambridge, Mass: Harvard University Press.

Hart, R. and Moore, G. (1973) 'The Development of Spatial Cognition: A Review', in R. Downs and D. Stea (eds) *Image and Environment: Cognitive Mapping and Spatial Behaviour*, London: Edward Arnold.

Hartley, J.B. and Woodward D. (eds) (1987) *The History of Cartography (Vol.1): Cartography in Prehistoric, Ancient, and Medieval Europe and the Mediterranean*, Chicago: University of Chicago Press.

Harvey, D. (2007 [1991]) *The Condition of Postmodernism: An Enquiry into the Origins of Cultural Change*, Oxford: Wiley-Blackwell.

Haselberger, L. (2009) 'The Pantheon: Nagging Questions to No End', in G. Graßhoff, M. Heinzelmann and M. Wafler (eds) *The Pantheon in Rome: Contributions*, Bern: Redaktion: Bernd Kulawik – Jon Albers, pp. 171–186.

Heather, P. (2009) *Empires and Barbarians: Migration, Development and the Birth of Europe*, London: Pan Books.

Heidegger, M. (1997a) 'Building, Dwelling, Thinking', in N. Leach (ed.) *Rethinking Architecture: A Reader in Architectural Theory*, London and New York: Routledge, 100–109.

Heidegger, M. (1997b) 'Techne', in N. Leach (ed.) *Rethinking Architecture: A Reader in Architectural Theory*, London and New York: Routledge, 120–121.

Hein, H. S. (2000) *The Museum in Transition: A Philosophical Perspective*, Washington, DC: Smithsonian Institution.

Hendrix, J. (2006) *Architecture and Psychoanalysis*, New York: Peter Lang.

Hendry, J. (2003 [1987]) *Understanding Japanese Society*, London: Routledge.

Herrle, P. (2008) 'Architecture and Identity? Steppenwolf and the Carriers of Change', in P. Herrle and E. Wegerhoff (eds.) *Architecture and Identity*, Berlin: Lit Verlag Dr W. Hopf.

Herrle, P. and Schmitz, S. (2009) 'Introduction', in *Constructing Identities in Contemporary Architecture: Case Studies from the South*, Berlin: Lit Verlag Dr W. Hopf.

Heslin, P. J. (2007) 'Augustus, Domitian and the so-called Horologium- Augusti', *Journal of Roman Studies*, vol. 97: 1–20.

Hester, R. T. Jr. (1993) 'Sacred Structures and Everyday Life: A Return to Manteo, North Carolina', in D. Seamon (ed.) *Dwelling, Seeing, and Designing: Towards a Phenomenological Ecology*, Albany: State University of New York Press.

Heynen, H. (1999) *Architecture and Modernity: A Critique*, Cambridge, Mass: MIT Press.

High Court of Australia (1992) *Mabo V Queensland (No 2) [1992] HCA 23; (1992) 175 CLR 1 (3 June 1992)*. Available at http://www.austlii.edu.au/cgi-bin/sinodisp/au/cases/cth/high_ct/175clr1.html?stem=0&synonyms=0&query=~mabo (accessed 3 June, 2012).

Hillen, S. (2007) *Constant Light* Available at http://www.seanhillen.com/Omagh2.html

Hillier, J. and Rooksby, E. (2005) 'Introduction to Second Edition: Committed Scholarship', in J. Hillier and E. Rooksby (eds) *Habitus: A Sense of Place*, Aldershot: Ashgate.

Hilton, C. (1957) *Be My Guest*, New Jersey: Prentice Hall.

Hitchcock, H. R. (1955) *Latin American Architecture since 1945*, New York: The Museum of Modern Art.

HMI (Henry Moore Institute), Helen Chadwick Collection. The text is cited verbatim and is referenced according to its archive accession number; the last two digits give the sketchbook number and page(s).

Hobsbawn, E. J. (2003 [1990]) *Nations and Nationalism since 1780: Programme, Myth, Reality*, Cambridge: Cambridge University Press.

Holl, S., Pallasmaa, J. and Pérez-Gómez, A. (2006) *Questions of Perception: Phenomenology of Architecture*, San Francisco: William K. Stout.

Horden, P. and Purcell, N. (eds) (2010) *The Corrupting Sea: A Study of Mediterranean History*, Oxford: Blackwell Publishing.

Hornblower, S. and Spawforth A. (eds) (1996) *The Oxford Classical Dictionary*, 3rd edition, Oxford: Oxford University Press.

House and Garden (1931) 'Modernist Houses in Mexico', 80, pp. 56–57.

House and Garden (1937) 'Modern in Mexico', 72, pp. 40–41.

Hunt, P. (2007) 'Imperium in the Pantheon and its Pavimentum' (*Philolog – Classical Connections, commentary and critique*). Available at http://traumwerk.stanford.edu/philolog/2005/12/imperium_in_the_pantheon_in_ro.html

Huyssen, A. (2003) *Present Pasts Urban Palimpsests and the Politics of Memory*, Stanford, California: Stanford University Press.

Ilardi, V. (1956) '"Italianita" among some Italian Intellectuals in the Early Sixteenth Century', *Traditio*, 12: 339–67.

INBA, *Ex templo de Santa Teresa la antigua/ Ex Teresa Arte Actual*, Instituto Nacional Bellas Artes: Museos. Available at http://exteresa.blogspot.co.uk/ (accessed 30 April 2009).

Ingold, T. (2000) *Perception of the Environment: Essays on Livelihood, Dwelling and Skill*, London: Routledge.

Jameson, F. (1991) *Postmodernism – or, The Cultural Logic of Late Capitalism*, London: Verso.

Jenkins, R. (2002) *Pierre Bourdieu*, London: Routledge.

Jodido, P. (1997) *Contemporary Japanese Architects Volume 2*, Koln: Taschen.

Jones, P. (2007) *Ochre and Rust: Artefacts and Encounters on Australian Frontiers*, Wakefield, SA: Wakefield Press.

Jones, W.R. (1971) 'The Image of the Barbarian in Medieval Europe', *Comparative Studies in Society and History*, vol. 13, Issue 04: 376–407.

Jordan, J. (2006) *Understanding Urban Change in Berlin and Beyond*, Stanford: Stanford University Press.

Journal of the Institute of Architects of Malaya (JIAM): 1(1–6, Jun–Nov 1931); 2(1–6, Dec–May 1931–32); 3(1–6, Jun–Nov 1932); 4(1–7, Dec–Jun 1932–33).

Journal of the Singapore Society of Architects (JSSA): 1 (1–12, Dec 1923–Nov 1924); 2(1–8, Jan–Aug 1925).

Journal of the Singapore Institute of Architects (SIAJ): 1–188.

Journal of the Singapore Society of Architects Incorporated (JSSAI): 1(1–6, Jun–Nov 1929); 2(1–6, Dec-May 1929–30); 3(1–6, Jun–Nov 1930); 4(1–6, Dec–May 1930–31).

Journal of the Society of Malayan Architects (JSMA): 1(1, 1958).

Judt, T. (2011) *The Memory Chalet*, New York: Penguin.

Kandel, E. (2007) *In Search of Memory: The Emergence of a New Science of Mind*, New York: Norton.

Kattago, S. (2001) *Ambiguous Memory: The Nazi Past and German National Identity*, Westport, CT: Praeger Publishers.

Kawano, S. (2005) *Ritual Practice in Modern Japan: Ordering People, Place and Action*, Honolulu: University of Hawaii Press.

Kearney, R. (1997) *Postnationalist Ireland: Politics Culture Philosophy*, London: Routledge.

Kelly, A. (2005) 'Geopolitical Eclipse: Culture and the Peace Process in Northern Ireland', *Third Text*, vol. 19, Issue 5, pp. 545–553.

King, A. (1990) 'Architecture, Capital and the Globalization of Culture', in M. Featherstone (ed.) *Global Culture: Nationalism, Globalization and Modernity*, London: Sage in association with Theory, Culture and Society, pp. 397–411.

King, A. (ed.) (1991) *Culture, Globalization, and the World System*, London: Macmillan.

Kłoczowski, J. (2000) *A History of Polish Christianity*, Cambridge: Cambridge University Press.

Knabe, C. and Noenning, J. (1999) *Shaking the Foundations: Japanese Architects in Dialogue*, Munich: Prestel.

Koestler, A. (1967) *The Ghost in the Machine*, London: Pan Piper/Hutchinson Publishing Group.

Koestler, A. (1982 [1959]) *The Sleepwalkers: A History of Man's Changing Vision of the Universe*, London: Penguin.

Kojin, K. (1980) *Nihon kindai bungaku no kigen*, Tokyo: Kodansha.

Kossowski, A. (1933) *Protestantyzm w Lublinie i w Lubelskiem w XVI–XVII w.*, Lublin: Dom Książki Polskiej.

Kuma, K. (2000) 'Dissolution of Objects and Evasion of the City', *Japan Architect*, vol. 38 (Summer), p. 58.

Kungel, R.M. (2000) 'Późnogotyckie cerkwie Wielkiego Księstwa Litewskiego,' in J. Lilejko (ed.) *Sztuka ziem wschodnich Rzeczypospolitej XVI-XVIII w.*, Lublin: Towarzystwo Naukowe KUL.

Kwint, M. *et al.* (1995) *Material Memories: Design and Evocation*, Oxford: Berg.

Lacan, J. (1968 [1956]) *The Language of the Self: The Function of Language in Psychoanalysis*, trans. A. Wilden, Baltimore and London: The Johns Hopkins University Press.

Lacan, J. (1977 [1966]) *Écrits, A Selection*, trans. A. Sheridan, New York: W.W. Norton.

Lacan, J. (1981 [1973]) *The Four Fundamental Concepts of Psycho-Analysis*, J.A. Miller (ed.), trans. A. Sheridan, New York: W.W. Norton.

Lai, C. (2005) 'Concrete/Concentric Nationalism: The Architecture of Independence in Malaysia, 1945–1969', unpublished PhD thesis, University of California, Berkeley.

Landsberg, A. (2004) *Prosthetic Memory: The Transformation of American Remembrance in the Age of Mass Culture*, New York: Columbia University Press.

Le Corbusier (1991) 'Brazilian Corollary . . . Which is also Uruguay', in *Precision: On the Present State of Architecture and City Planning*, Cambridge, Mass: MIT Press, pp. 233–245.

Le Corbusier (1995) '1936–1945, Les Ministère de l'Education nationale el de la Santé publique à Rio de Janeiro', in *Œevre Complèt 4, 1938–1946*, Zurich: Les Editions d'Architecture, pp. 80–89.

Leach, N. (2000) 'Virilio and Architecture', in J. Armitage (ed.) *Paul Virilio: From Modernism to Hypermodernism and Beyond*, London: Sage Publications.

Leach, N. (2005 [2002]) 'Belonging: Towards a Theory of Identification with Space', in J. Hillier and E. Rooksby (eds) *Habitus: A Sense of Place*, Aldershot and Farnham: Ashgate Publishing Ltd.

Leach, N. (2006) *Camouflage*, Cambridge, Mass: MIT Press.

Leatherbarrow, D. (2002) *Uncommon Ground: Architecture, Technology, and Topography*, Cambridge, Mass. and London: MIT Press.

Lefaivre, L. and Tzonis, A. (2001) *Tropical Architecture: Critical Regionalism in the Age of Globalization*, London: Wiley-Academy.

Lefaivre, L. and Tzonis, A. (2012) *Architecture of Regionalism in the Age of Globalization: Peaks and Valleys in the Flat World*, New York and London: Routledge.

Lefebvre, H. (1991) *The Production of Space* D. Nicholson-Smith (trans.), Oxford and Cambridge, Mass: Blackwell Publishing.

Lekan, T. M. (2004) *Imagining the Nation in Nature: Landscape Preservation and German Identity, 1885–1945*, Cambridge, Mass: Harvard University Press.

Liernur, J. F. (2002) *Escritos de arquitectura del siglo 20 en América Latina*, Madrid: Tanais Ediciones.

Liturgical Arts Vol. 21–Vol. 22 (1953), series titled 'Around the World in Ninety Days'.

Longley, E. (2001) 'Northern Ireland: Commemoration, Elegy, Forgetting', in I. McBride (ed.) *History and Memory in Modern Ireland*, Cambridge University Press.

Loomba, A. and Kaul, S. (1994) 'Introduction: Location, Culture and Post-Coloniality', in *On India: Writing History, Culture and Post-Coloniality, Oxford Literary Review* 16: 3–30.

Lovell, N. (1998) 'Introduction', in N. Lovell (ed.) *Locality and Belonging*, London: Routledge.

Lozoya, J. (2010) *Las manos indígenas de la raza española. El mestizaje como argumento arquitectónico*, Mexico City: CONACULTA.

MacDonald, S. (2006) 'Undesirable Heritage: Fascist Material Culture and Historical Consciousness in Nuremberg', *International Journal of Heritage Studies*, 12, 1: 9–28.

MacDonald, W. L. (1976) *The Pantheon: Design, Meaning, and Progeny*, London: Allen Lane.

Maier, C. S. (1997 [1988]) *The Unmasterable Past: History, Holocaust and German National Identity*, Cambridge, Mass: Harvard University Press.

Maldonado, A.W. (1997) *Teodoro Moscoso and Puerto Rico's Operation Bootstrap*, Gainsville: University Press of Florida.

Marqués Mera, J. (1998) 'Entrevista al arquitecto Miguel Ferrer, FAIA', unpublished interview (8 September 1998).

Martinez, D. P. (2005) 'On the "Nature" of Japanese Culture, or, Is There a Japanese Sense of Nature?', in J. Robertson (ed.) *A Companion to the Anthropology of Japan*, Malden: Blackwell.

Martocci, F. T. (1947) *Promotional Tourist Hotel*, Henry Klumb Collection, box 83.34, Architecture and Construction Archives at the University of Puerto Rico, Río Piedras.

Massey, D. (1994) *Space, Place and Gender*, Cambridge: Polity Press.

Matero, F. (2000) 'Ethics and Policy in Conservation', *Conservation: The Getty Conservation Institute Newsletter*, 15: 5–9.

Mathews, G. (2000) *Global Culture / Individual Identity*, London: Routledge.

McCrone, D. (1992) *Understanding Scotland: The Sociology of a Stateless Nation*, London: Routledge.

McQuire, P. (2000) 'Blinded by the Speed of Light', in J. Armitage (ed.) *Paul Virilio: From Modernism to Hypermodernism and Beyond*, London: Sage Publications.

Mendez-Vigata, A. E. (1997) 'Politics and Architectural Language: Post-Revolutionary Regimes in Mexico and Their Influence on Mexican Public Architecture, 1920–1952', in E. Burian (ed.) *Modernity and the Architecture of Mexico*, Austin, Texas: University of Texas Press, pp. 61–90.

Merleau-Ponty, M. (2002 [1962]) *The Phenomenology of Perception*, trans. C. Smith, London, New York: Routledge.

Messent, D. (1997) *Opera House Act One*, Sydney: David Messent Photography.

Mignolo, W. D. (2007) *La Idea de América Latina: La Herida Colonial y la Opción Decolonial*, trans. S. Jawerbaum and J. Barba, Barcelona: Gedisa Editorial.

Misztal, B. A. (2003) *Theories of Social Remembering*, Philadelphia: Open University Press.

Mitchell, T. (2002) 'McJihad: Islam in the U.S. Global Order', *Social Text*, *20* (4), pp. 1–18.

Mohanty, S.P. (2000) 'The Epistemic Status of Cultural Identity: On *Beloved* and the Postcolonial Condition', in P. M. L. Moya and M. R. Hames-García (eds) *Reclaiming Identity:*

Realist Theory and the Predicament of Postmodernism, Berkeley: University of California Press.

Monetary Authority of Singapore (MAS) *Currency Interchangeability*. Available at www.mas.gov.sg/currency/currency_info/Currency_Interchangeability_Agreement.html (accessed 30 Jan 2012).

Monetary Authority of Singapore (MAS) *Heritage Collection: The Currency History of Singapore*. Available at www.mas.gov.sg/currency/currency_info/Heritage_Collection.html (accessed 30 Jan 2012).

Moraña, M., Dussel, E. and Jáuregui, C. (eds) (2008) *Coloniality at Large: Latin America and the Postcolonial Debate*, Durham and London: Duke University Press.

Morrow, E. (1932) *Casa Mañana*, New York: The Spiral Press.

Moscoso, T. (1954) 'El desarrollo económico de Puerto Rico: Problemas de Puerto Rico en relación con su población y sus recursos', *Conferencias a las Naciones Unidas, Comisión Económica para la América Latina, Programa de Capacitación*, Santiago de Chile, Teodoro Moscoso Collection, Luis Muñoz Marín Foundation Archives, San Juan, Puerto Rico.

Mostafavi, M, and Leatherbarrow, D. (1993) *On Weathering: Life of Buildings in Time*, Cambridge, Mass: MIT Press.

Moya, P.M.L., and Hames-Garcia, M.R. (2000) *Reclaiming Identity: Realist Theory and the Predicament of Postmodernism*, Berkeley: University of California Press.

Muñoz Marín, L. (1959) 'Breakthrough from Nationalism: A Small Island Looks at a Big Trouble', *Godkin Lecture* at Harvard University (28 April 1959), cited in P. Reina Pérez (2005) *Cavilando el Fin del Mundo: Apología y Confesión en las Conferencias Godkin 1959 de Luis Muñoz Marín*, San Juan: West Álamo Caribbean Publishing, pp. 131–141.

Muñoz Marín, L. (1992) *Memorias: 1940–1952*, San Germán: Universidad Interamericana de Puerto Rico.

Murtagh, B. (2008) 'New space and Old in "Post Conflict" Belfast', *Divided Cities/Contested States*. Working Paper no. 5, Queen's University Belfast: Belfast. Available at www.arct.cam.ac.uk/conflictincities/workingpapers.html (accessed January 2009).

Musil, R. (2006) *Posthumous Papers of a Living Author*, New York: Archipelago Books.

Musteikis, A. (1988) *The Reformation in Lithuania: Religious Fluctuations in the Sixteenth Century*, Boulder, CO: East European Monographs, Columbia University Press.

Myers, I.E. (1952) *Mexico's Modern Architecture*, New York: Architectural Book Publishing Company.

Nast, H. and Pile, S. (1998) 'Everydayplacesbodies', in H. Nast and S. Pile (eds) *Places through the Body*, London: Routledge.

National Archives of Singapore, Building Plans 633A/38 dated June 1939.

Neutra, R. (1952) 'Introduction', in I. E. Myers (1952) *Mexico's Modern Architecture*, New York: Architectural Book Publishing Company, pp. 20–22.

Nitschke, G. (1993) '*En*: Transactional Space', in *From Shinto to Ando*, London: Academy Group, pp. 85–93.

Noelle, L. (2004) 'Estados Unidos y la arquitectura mexicana en el siglo XX: El punto de vista de las publicaciones', *Anales del Instituto de Investigaciones Estéticas*, UNAM, 85: 49–60.

Noever, P., Witt-Dörring, C. and Dagdelen, C. (eds) (2004) *Canan Dagdelen: Yurt Tutmusch Dot*, Vienna: MAK, Austrian Museum of Applied Art.

Noland, C. (2009) *Agency and Embodiment: Performing Gestures/Producing Culture*, Cambridge, Mass: Harvard University Press.

Nora, P. (2002) 'Reasons for the Upsurge in Memory', *Eurozine*. Available at www.eurozine.com/articles/2002-04-19-nora-en.html (accessed February 2008).

O'Beirn, A. (2006) *Space Shuttle*. Available at www.spaceshuttle.org (accessed January 2009).

O'Beirn, A. (2008) *Arkiv City*. Available at http://interface.ulster.ac.uk/arkivecity/aisling_O_beirne.html (accessed January 2009).

O'Connell, A. *Inoculacion*, Interview via email. 22 August 2009.

O´Gorman, P. (1980) *Tradition of Craftsmanship in Mexican Homes*, New York: Architectural Book Publishing Co.

O´Gorman, P. (1988) *Patios and Gardens of Mexico*, Lanham: Taylor Trade Publishing.

Okabe, N. (2000) 'Slow Transition', trans. A. Barrrie and A. Wakasugi, *JA Yearbook 2000*.

Pamuk, O. (2009) *The Museum of Innocence*, trans. M. Freely, New York: Alfred A. Knopf.

Parkes, G. (1995 [1993]) 'Ways of Japanese Thinking', in N. Hume (ed.) *Japanese Aesthetics and Culture*, Albany: State University of New York Press.

Paz, O. (1985) *The Labyrinth of Solitude and Other Writings*, New York: Grove Press.

Pérez Gómez, A. (1988) 'Abstraction in Modern Architecture: Some Reflections in Parallel to Gnosticism and Hermeneutics', in *VIA: Journal of the Graduate School of Fine Arts*, University of Pennsylvania, 71–83.

Pérez Gómez, A. (2006) *Built Upon Love: Architectural Longing After Ethics and Aesthetics*, Cambridge, Mass: MIT Press.

PETA: Journal of the Federation of Malaya Society of Architects 1(1), June 1955.

Phillip, A. (1788) *Letter: Phillip to Banks*, November 16, C.213. Sydney: Mitchell Library.

Phillips, A. (2007) *Side Effects*, London: Penguin Books.

Piotrowski, A. (2011) *Architecture of Thought*, Minneapolis: University of Minnesota Press.

Poignant, R. (2004) *Professional Savages: Captive Lives and Western Spectacles*, New Haven, CT: Yale University Press.

Pratt, M. L. (1992) *Imperial Eyes: Travel Writing and Transculturation*, London: Routledge.

Pred, A. (1984) 'Place as Historically Contingent Process: Structuration and the Time-Geography of Becoming Places', *Annals of the Association of American Geographers*, 74(2): 279–297.

Quek, R. (2012a) 'Editorial Commentary: Architecture and the Construction of Identity', in *National Identities*, Vol. 14, no. 2, Routledge.

Quek, R. (2012b) 'Dramatic Pre-emption in Singapore's National Theatre: Constructing National Identity before an Independent Nation', in *National Identities*, Vol. 14, no. 2, Routledge.

Radstone, S. (ed.) (2000) *Memory and Methodology*, Oxford: Berg.

Reader, I. (1991) *Religion in Contemporary Japan*, Honolulu: University of Hawaii.

Reina Pérez, P. (2003) *La Semilla que Sembramos: Autobiografía del Proyecto Nacional*, San Juan: Editorial Cultural.

Reina Pérez, P. (2005) *Cavilando el Fin del Mundo: Apología y Confesión en las Conferencias Godkin 1959 de Luis Muñoz Marín*, San Juan: West Álamo Caribbean Publishing.

Reyes Padró, C. (1949) 'Cientos de Invitados Asistieron a la Inauguración del Hilton', *El Mundo*, San Juan (12 December): 1, 16.

Riegl, A. (1982 [1903]) 'The Modern Cult of Monuments: Its Character and Its Origin', *Oppositions*, 25: 21–51, Cambridge, Mass: MIT Press.

Ríos Garza, C. (ed.) (2004a) *Revista El Arquitecto 1923–1927: Estudio introductorio y análisis del contenido*, Mexico City: Universidad Nacional Autónoma de México.

Ríos Garza, C. (ed.) (2004b) *Arquitectura y lo demás: 1945–1950. Estudio introductorio y análisis de contenido*, Mexico City: Universidad Autónoma de México.

Ríos Garza, C. (ed.) and Noelle, L. (introductory study) (2008) *Revista Arquitectura México. 119 números. 1938–1978*, Mexico City: Universidad Autónoma de México.

Ríos Garza, C., *et al.* (eds) (2001 [1934]) *Pláticas sobre arquitectura. México, 1933*, Mexico City: UNAM and UAM-Azcapotzalco.

Robertson, R. (1991) 'Social Theory, Cultural Relativity, and the Problems of Globality', in A. King (ed.) *Culture, Globalization, and the World System*, London: Macmillan, pp. 60–90.

Roderick, L. (1995) 'Deciphering Home: An Integrative Historical Perspective', in D. Benjamin (ed.) *The Home: Words, Interpretations, Meanings and Environments*, Avebury: Aldershot.

Ross, D. F. (1966) *The Long Uphill Path: A Historical Study of Puerto Rico's Progress of Economic Development*, San Juan: Talleres Gráficos Interamericanos.

Rossbach, S. (2007) 'The Impact of 'exile' on thought: Plotinus, Derrida and Gnosticism', *History of the Human Sciences*, vol. 20, no. 4: 27–52. Available at http://hhs.sagepub.com/cgi/reprint/20/4/27.

Rossell, L. E. (1979) *Iglesias y Conventos Coloniales de México*, México: Editorial Patria.

Rossi, L. (1971) *Trajan's Column and the Dacian Wars*, trans. J.M.C. Toynbee, Ithaca: Cornell University Press.

Rothberg, M. (2009) *Multidirectional Memory: Remembering the Holocaust in the Age of Decolonization*, Stanford, California: Stanford University Press.

Rowe, C. and Slutzky, R. (1963) 'Transparency: Literal and Phenomenal', in C. Rowe (1976) *The Mathematics of the Ideal Villa and Other Essays*, Cambridge, Mass: MIT Press.

Różycka-Bryzek, A. (1992) 'Bizantyńsko-słowiańskie malowidła w gotyckich kościołach Polski pierwszych Jagiellonów', in T. Chrzanowski (ed.) *Między Wschodem i Zachodem, Część III, Kultura artystyczna*, Lublin: Lubelskie Towarzystwo Naukowe.

Różycka-Bryzek, A. (2000) *Freski w kaplicy zamku lubelskiego*, Lublin: Maria Curie-Skłodowka University Press.

Rumah: *Journal of the Society of Malayan Architects* (*RJSMA*): 2–3(1959–60).

Rumah: *Journal of The Singapore Institute of Architects* (*RJSIA*): 4–9(1961–69).

Rushdie, S. (1988) *The Satanic Verses*, New York: St. Martin's Press.

Ruskin, J. (1989) *The Seven Lamps of Architecture*, New York: Dover Publications.

Rutherford, J. (1990) 'The Third Space: Interview with Homi Bhabha', in J. Rutherford (ed.) *Identity: Community, Culture, Difference*, London: Lawrence and Wishart.

Rykwert, J. (2004) *The Seduction of Place: The City in the Twenty-First Century and Beyond*, London: Weidenfeld and Nicolson.

Said, E. (1979). *Orientalism*. New York: Vintage Books.

Said, E. (1993) *Culture and Imperialism*, New York: Knopf, London: Vantage Books.

Said, E. (2004) *Humanism and Democratic Criticism*. New York: Columbia University Press.

Salmon, E. T. (1936) 'Trajan's Conquest of Dacia', *Transactions and Proceedings of the American Philological Association*, vol. 67: 83–105.

Salvatore, R. D. (2005) 'Panamericanismo práctico: Acerca de la mecánica de la penetración comercial norteamericana', in *Culturas Imperiales: Experiencia y Representación en América, Asia y África*, Argentina: Beatriz Viterbo Editora, pp. 271–300.

Samsonowicz, H. (ed.) (1984) *Polska Jana Długosza*, Warszawa: Państwowe Wydawnictwo Naukowe.

Santino, J. (2006) *Spontaneous Shrines and the Public Memorialisation of Death*, New York: Palgrave-Macmillan.

Sarup, M. (1996) *Identity, Culture and The Postmodern World*, Edinburgh: Edinburgh University Press.

Saussure, F. de (1986 [1916]) *Course in General Linguistics*, C. Bally, A. Sechehaye and A. Riedlinger (eds.), trans. R. Harris, Chicago: Open Court.

Schlink, B. (2010) *Guilt about the Past*, London: Beautiful.

Sklovsky, V. (1917) 'Art as Technique', in L. T. Lemon and M. J. Reiss (eds) (1965) *Russian Formalist Criticism: Four Essays*, Lincoln: University of Nebraska Press, pp. 3–24.

Sennett, R. (2002) 'The Foreigner', in G. Dodds and R. Tavernor (eds.) *Body and Building: Essays on the Changing Relation of Body and Architecture*, Cambridge, Mass: MIT Press.

Sharr, A. (ed.) (2012) *Reading Architecture and Culture: Researching Buildings, Spaces, and Documents*, London, New York: Routledge.

Shinohara, K. (1982) 'Shinohara', *IAUS 17*, Rizzoli: New York.

Shipway, V. C. and Shipway, W. (1960) *The Mexican House Old and New*, New York: Architectural Book Publishing Co.

Shipway, V. C. and Shipway, W. (1962) *Mexican Interiors*, New York: Architectural Book Publishing Co.

Shipway, V. C. and Shipway, W. (1964) *Mexican Homes of Today*, New York: Architectural Book Publishing Co.

Shipway, V. C. and Shipway, W. (1966) *Decorative Design in Mexican Homes*, New York: Architectural Book Publishing Co.

Shipway, V. C. and Shipway, W. (1970) *Houses of Mexico: Origins and Traditions*, New York: Architectural Book Publishing Co.

Singapore Architect (SA): 189–present.

Smith, C. B. (1967) *Builders in the Sun: Five Mexican Architects*, New York: Architectural Book Publishing Co.

Smith, K.V. (2001) *Bennelong: The Coming in of the Eora: Sydney Cove 1788–1792*, East Roseville, NSW: Kangaroo Press.

Smith, T. (2006) *The Architecture of Aftermath*, Chicago: University of Chicago Press.

Snyder, L. (1954) *The Meaning of Nationalism*, Westport: Greenwood Press.

Souter, G. (1973) 'Where Palms Once Waved', *Sydney Morning Herald*, 15 October.

Spivak, G. C. (1987) *In Other Worlds: Essays in Cultural Politics*, New York and London: Methuen.

Spratling, W. (1927) 'Some Impressions of Mexico', *Architectural Forum 47*, pp. 1–8 and pp. 161–168.

Spratling, W. (1929) 'Indo-Hispanica Mexico', *Architecture 59*, pp. 75–144.

Spratling, W. (1932) *Little Mexico*, New York: Jonathan Cape and Harrison Smith.

Stagno, B. (2001) 'Tropicality', in A. Tzonis, L. Lefaivre and B. Stagno (eds) *Tropical Architecture: Critical Regionalism in the Age of Globalization*, Chichester: Wiley Academy, pp. 65–92.

Statistics Bureau, Ministry of Internal Affairs and Communications (2011) 'Historical Statistics of Japan', Available at www.stat.go.jp/english/data/chouki/02.htm (accessed 20 December 2011).

Stewart, D. (1987) *The Making of Modern Japanese Architecture: 1968 to the Present*, Tokyo: Kodansha.

Street-Porter, T. (1989) *Casa Mexicana: The Architecture, Design, and Style of Mexico*, New York: Stewart, Tabori and Chan.

Sturken, M. (2007) *Tourists of History: Memory, Kitsch and Consumerism from Oklahoma City to Ground Zero*, Durham NC: Duke University Press.

Tacitus, A. (1999) *Agricola and Germany*, trans. A.R. Birley, Oxford: Oxford University Press.

Taki, K. (1988) 'Fragments and Noise: The Architectural Ideas of Kazuo Shinohara and Toyo Ito', trans. A. Birnbaum, in B. Bognar (ed.) *Japanese Architecture*, London: Academy Group.

Talbert, R. J. A. (2010) *Rome's World: The Peutinger Map Reconsidered*, Cambridge: Cambridge University Press.

Tanaka, S. and Tanaka, S. (1998) *The Tea Ceremony*, Tokyo: Kodansha.

Taussig, M. (2004) 'Jimmie Durham' in I. Carlos (*ed.*) *On Reason and Emotion: Biennale of Sydney 2004*, Sydney: Art Gallery of NSW.

Taylor, C. (1999) 'To Follow a Rule . . .', in R. Shusterman (ed.) *Bourdieu: A Critical Reader*, Oxford: Blackwell.

Temple, N. and Bandyopadhyay, S. (2007) 'Contemplating the Unfinished: Architectural Drawing and the Fabricated Ruin', in M. Frascari, *et al* (eds.) *From Models to Drawings: Imagination and Representation in Architecture*, London: Routledge.

Tench, W. (1793) with contrib. L.F. Fitzhardinge (1961) *Sydney's First Four Years: Being a Reprint of a Narrative of the Expedition to Botany Bay and a Complete Account of the Settlement at Port Jackson*, Sydney: Angus and Robertson in association with the Royal Australian Historical Society.

The Architectural Record (1931) 'Mexican Villas', 69, pp. 162–64.

The Architectural Record (1935a) 'Recent Work of a Mexican Architect: Luis Barragán', 77, pp. 33–46.

The Architectural Record (1935b) 'Children's Park: Parque de la Revolución Guadalajara Jalisco, México', 78, pp. 165–69.

The Architectural Record (1936) 'School-Home for Small Children: José V. García and Enrique de la Mora', 79, pp. 445–50.

The Architectural Record (1944) 'Children's Hospital. Mexico City', 96, pp. 60–7.

The Malayan Architect (*TMA*): 5(1, Jul 1933; 6, Dec 1933); 6–12(1–12, Jan–Dec 1934–40); 13(1–11, Jan–Nov 1941).

The Quarterly Journal of the Institute of Architects of Malaya (*QJIAM*): 1–6 (Mar 1951–1958).

The Sydney Opera House Trust (1973) 'Opening Ceremony Souvenir' in *Ephemera Relating to the Sydney Opera House: Describing the Planning, Design and Development of the Building*, 1950, Sydney: State Library of NSW.

The Sydney Opera House Trust (2006) 'House History'. Available at www.sydneyoperahouse.com/about/house_history/1788_1953.aspx (accessed 3 January 2012).

TheSydneyOperaHouseTrust(2006)'HouseHistory'.Availableatwww.sydneyoperahouse.com/sections/about_the_house?house_history/?sm=4andss=18 (accessed 26 May 2006).

The Sydney Opera House Trust (2012) 'House History'. Available at www.sydneyoperahouse.com/about/house_history/1788_1953.aspx (accessed 3 January 2012).

Theocharis, N. (2009) 'How the Bern Digital Pantheon Model Can Help Answer <Nagging Questions> about the Pantheon', in G. Graßhoff, M. Heinzelmann and M. Wafler (eds) *The Pantheon in Rome: Contributions*, Bern: Redaktion: Bernd Kulawik – Jon Albers pp. 187–93.

Thomas, N. (1991) *Entangled Objects: Exchange, Material Culture, and Colonialism in the Pacific*, Cambridge, Mass: Harvard University Press.

Time Magazine (1958) 'Democracy's Laboratory in Latin America', 71/25 (June 23 1958): cover.

Tonkiss, F. (2005) *Space, the City and Social Theory*, Cambridge: Polity Press.

Torisson, F. (2010) *Berlin: Matter of Memory*, London: Ratatosk Publishing Ltd.

Torres Santiago, J. (2000) 'The Invention of the Gates of Paradise: Images, Architecture, and Context in the Development of Hotels in San Juan', trans. A. Hurley, in E V. Farage (ed.) (2000) *Ever New San Juan: Architecture and Modernization in the Twentieth Century*, San Juan: San Juan 2000 Commission and AACUPR, pp. 118–67.

Trajdos, T. M. (1979) 'Od lubelskiej kaplicy do politycznych planów Jagiełły', *Spotkania z zabytkami*, R. 2.

Tuan, Y. (1991 [1977]) *Space and Place: The Perspective of Experience*, Minneapolis: University of Minnesota Press.

Tugwell, R.G. (1947) *The Stricken Land: The Story of Puerto Rico*, New York: Doubleday and Company.

Tzonis. A. and Lefaivre, L. (1981) 'The Grid and the Pathway. An Introduction to the Work of Dimitris and Suzana Antonakakis', in *Architecture in Greece* 15, Athens.

Vale, L. (1992) *Architecture Power and National Identity*, New Haven: Yale University Press.

Véjar Pérez-Rubio, C. (2007) *La espiral del sincretismo: En busca de una identidad para nuestra arquitectura*, México: UNAM, UIA and UAM.

Vellinga, M. (2006) 'Engaging the Future: Vernacular Studies in the 21st Century', in L. Asquith and M. Vellinga (eds) *Vernacular Architecture in the Twenty-First Century: Theory, Education and Practice*, London: Routledge.

Verna Cook Shipway Papers, Mandeville Special Collections Library, Geisel Library, University of California, San Diego, MSS 0105.

Versnel, H.S. (1970) *Triumphus: An Enquiry into the Origin, Development and Meaning of the Roman Triumph*, Leiden: E.J. Brill.

Villagrán, J. (1967) 'Foreword', in C. Bamford Smith, *Builders in the Sun: Five Mexican Architects*, New York: Architectural Book Publishing Co., pp. 9–14

Vitruvius (1931 [27 BC]) *On Architecture*, trans. F. Granger, Cambridge, Mass: Harvard University Press.

Vivoni Farage, E. (1995) 'Hacia una Modernidad Tropical: La Obra de Henry Klumb, 1928–1984', *Capitel: Boletin Informativo del Colegio de Arquitectos de Puerto Rico*, May: 6–11.

Voegelin, E. (1974) *Order and History (Volume Four): The Ecumenic Age*, Baton Rouge: Louisiana State University Press.

von Hildebrand, A. (1893) 'The Problem of Form in the Fine Arts', trans. H. F. Mallgrave and E. Ikonomou (1994), in *Empathy, Form and Space: Problems in German Aesthetics 1873–93*, Los Angeles: Getty Center for the History of Art and the Humanities, pp. 227–280.

Walicki, M. (1930) 'Malowidła ścienne kościoła Św. Trójcy na Zamku w Lublinie', *Studia do dziejów Sztuki w Polsce*, t. 3.

Walker, S. (2013) *Helen Chadwick: Constructing Identities Between Art and Architecture*, London: I. B. Tauris.

Wallerstein, I. (1991) 'The Nation and the Universal: Can There Be Such a Thing as a World Culture?', in A. King (ed.) *Culture, Globalization, and the World System*, London: Macmillan, pp. 91–106.

Walsh, K. (1992 [1984]) *The Representation of the Past: Museums and Heritage in the Postmodern World*, London, New York: Routledge.

Weiner, A. (1985) 'Inalienable Wealth', *American Ethnologist* 12, cited in N. Thomas (1991) *Entangled Objects: Exchange, Material Culture, and Colonialism in the Pacific*, Cambridge, Mass: Harvard University Press.

Weiner, A. (1992) *Inalienable Possessions: The Paradox of Keeping-While-Giving*, Berkeley: University of California Press.

Weston, R. (2002) *Utzon: Inspiration, Vision, Architecture*. Hellerup, Denmark: Edition Blondal.

Wharton, A. (2001) *Building the Cold War: Hilton International Hotels and Modern Architecture*, Chicago: University of Chicago Press.

Willey, K. (1979) *When the Sky Fell Down: The Destruction of the Tribes of the Sydney Region, 1788–1850's*, Sydney: Collins.

Williams, P. (2007) *Memorial Museums: The Global Rush to Commemorate Atrocities*, Oxford: Berg.

Wittgenstein, L. (1963) *Philosophical Investigations*, trans. G. Anscombe, Oxford: Blackwell.

Witynski, K. and Carr J. P. (1999) *The New Hacienda*, Salt Lake City: Gibbs-Smith Publishers.

Wolff, J. (1991) 'The Global and the Specific: Reconsidering Conflicting Theories of Culture', in A. King (ed.) *Culture, Globalization, and the World System*, London: Macmillan, pp. 161–174.

Woodward, D. (ed.) (2007) *The History of Cartography, Volume 3: Cartography in the European Renaissance*, Chicago: University of Chicago Press.

Wright, G. (1991) *The Politics of Design in French Colonial Urbanism*, Chicago: University of Chicago Press.

Yánez, E. (1951) *18 Homes of Mexican Architects / 18 Residencias de arquitectos mexicanos*, Mexico City: Ediciones Mexicanas.

Yánez, E. (1952) 'Prólogo', in I. E. Myers (1952) *Mexico's Modern Architecture*, New York: Architectural Book Publishing Company.

Yatsuka, H. (1988) 'An Architecture Floating on the Sea of Signs: Three Generations of Contemporary Japanese Architects', in B. Bognar (ed.) *Japanese Architecture*, London: Academy Group.

Young, J. (1993) *The Texture of Memory: Holocaust Memorials and Meaning*, New Haven: Yale University Press.

Young, J. (2000) *At Memory's Edge: After-Images of the Holocaust in Contemporary Art and Architecture*, New Haven: Yale University Press.

Załęski, S. (1908) *Jezuici w Polsce: w skróceniu, 5 tomów w jednym, z dwoma mapami*, Kraków: Drukarnia W. L. Anczyca i Sp.

Index

Illustrations are shown by a page reference in italics.